The Biology and Medicine of Rabbits and Rodents

The Biology and Medicine of Rabbits and Rodents

JOHN E. HARKNESS, D.V.M., M.S., M.Ed.
Laboratory Animal Resources
The Pennsylvania State University
University Park, Pennsylvania

JOSEPH E. WAGNER, D.V.M., M.P.H., Ph.D.
College of Veterinary Medicine
University of Missouri-Columbia
Columbia, Missouri

Illustrations by DONALD L. CONNOR

LEA & FEBIGER  Philadelphia

Library of Congress Cataloging in Publication Data

Harkness, John E 1939
The biology and medicine of rabbits and rodents.

Includes bibliographies and index.
1. Laboratory animals—Diseases. I. Wagner, Joseph E., joint author. II. Title. [DNLM:
1. Animals, Laboratory.
QY50 H282L]
SF996.5.H37 619 76-23136
ISBN 0-8121-0576-1

PRINTED IN THE UNITED STATES OF AMERICA

Print number: 4 3

Preface

This book arose from the frustrations of repeated and abortive searches through numerous tomes in laboratory animal medicine for a practical, portable, accessible, clinically oriented text for the veterinarian, student, and biomedical investigator.

This text contains selected information on the biology, husbandry, therapy, clinical syndromes, and diseases of rabbits, guinea pigs, hamsters, gerbils, mice, and rats. The text provides a broad base of practical information; it is not a reference text, although the several bibliographies will facilitate in-depth examination of specific subject areas. The content, based on survey information, reflects the common husbandry and disease problems encountered in laboratory animals by veterinary practitioners, animal care personnel, and the outpatient clinics of several veterinary schools. The bibliographies in Chapter 4 will refer the reader to information about some of the rarer diseases or syndromes that are not discussed in detail in Chapter 5. Following the chapter on specific diseases, 40 actual cases are presented to indicate the complex and often baffling nature of clinical problems. Each chapter contains one or more bibliographies. For readers who wish information on the biology and diseases of nonhuman primates, a recent and concise text is the *Laboratory Primate Handbook* by R. A. Whitney, D. J. Johnson, and W. C. Cole (New York, Academic Press, Inc., 1973).

There is little new in this text, in substance or in form, except perhaps for the emphasis on clinical medicine, an emphasis which in some chapters may go beyond what is actually known of differential diagnoses, etiologies, and treatments in laboratory animals. We included these extrapolations not to mislead but to suggest alternatives that might direct investigators into areas requiring elucidation.

For the content and organization of the text we owe a considerable debt to the hundreds of authors and editors included in the bibliographies, but especially to Stephen M. Schuchman's Chapter "Individual Care and Treatment of Mice, Rats, Guinea Pigs, Hamsters, and Gerbils" in *Current Veterinary Therapy V* (R. W. Kirk, Editor, Philadelphia, W. B. Saunders Co., 1974) and to the Committee on Laboratory Animal Diseases, Chester A. Gleiser, Chairman, authors of *A Guide to Infectious Diseases of Guinea Pigs, Gerbils, Hamsters, and Rabbits* (Institute of

Laboratory Animal Resources, National Research Council, Washington, DC, 1974).

Portions of the sections on husbandry, clinical techniques, syndromes, and diseases were read by J. Darrell Clark, D.Sc., D.V.M., of the University of Georgia; Richard M. Hoar, Ph.D., of Hoffman-LaRoche, Inc.; Sigmund T. Rich, D.V.M., Consultant to the Salk Institute; Steven H. Weisbroth, D.V.M., M.S., of the State University of New York at Stony Brook; and Richard E. Doyle, D.V.M., M.S., Merlin C. Herrick, Ed.D., Lawrence G. Morehouse, D.V.M., Ph.D., Dwight R. Owens, M.S., and Charles E. Short, D.V.M., M.S., of the University of Missouri-Columbia. We did not always heed their admonitions, and errors remaining in fact and interpretation are ours alone. Where errors do occur, we would very much appreciate suggestions and corrections.

Portions of this text were developed at the Research Animal Diagnostic Laboratory, University of Missouri-Columbia. Development was supported in part by USPH Grants RR0471-07 and RR5006-08.

JOHN E. HARKNESS
JOSEPH E. WAGNER

Columbia, Missouri

Contents

CHAPTER 1 · GENERAL HUSBANDRY

Major Concerns in Husbandry 1
Factors Predisposing to Disease 2
Facility Cleaning and Disinfection 3
Formaldehyde Gas Fumigation 4
Bibliography 5

CHAPTER 2 · BIOLOGY AND HUSBANDRY

The Rabbit 7
The Guinea Pig 13
The Hamster 19
The Gerbil 24
The Mouse 29
The Rat 35

CHAPTER 3 · CLINICAL PROCEDURES

Drug Dosages and Therapeutic Regimens 41
Anesthetics and Anesthesia 46
Surgical Techniques 52
Radiography 53
Euthanasia 54
Serologic Testing for Rodent Viruses 54
Bibliography 55

CHAPTER 4 • CLINICAL SYNDROMES AND DIFFERENTIAL DIAGNOSES

History Protocol 59
Rabbit Syndromes 60
Guinea Pig Syndromes 62
Hamster Syndromes 64
Gerbil Syndromes 65
Mouse Syndromes 66
Rat Syndromes 68
Bibliography 69

CHAPTER 5 • SPECIFIC DISEASES

Acariasis 73
Bacillus piliformis Infection (Tyzzer's Disease) 76
Bordetella bronchiseptica Infection 77
Cestodiasis 78
Coccidiosis 80
Corynebacterium kutscheri Infection 82
Dermatophytosis 83
Ectromelia (Mouse Pox) 85
Encephalitozoonosis 86
Epizootic Diarrhea of Infant Mice (EDIM) 87
Heat Stroke 88
Hypovitaminosis C 89
Lymphocytic Choriomeningitis 90
Mouse Encephalomyelitis (Mouse Polio) 92
Mouse Hepatitis 93
Mucoid Enteropathy 94
Murine Chronic Mycoplasmosis 95
Neoplasia 96
Nephrosis 101
Pasteurella multocida Infection 102
Pasteurella pneumotropica Infection 104
Pediculosis 105
Pregnancy Toxemia 106
Proliferative Ileitis (Wet Tail) 107
Salmonellosis 109
Sendai Virus Infection 110
Streptococcus pneumoniae Infection 111

Streptococcus zooepidemicus Infection . . . 112
Tularemia . . . 114
Venereal Spirochetosis . . . 115
Miscellaneous Conditions . . . 115
Anorexia . . . 115
Dystocia . . . 116
Epileptiform Seizures in Gerbils . . . 116
Malocclusion . . . 116
Mastitis . . . 116
Moist Dermatitis . . . 117
Nitrate Toxicity . . . 117
Nonspecific Enteropathy . . . 117
Nutritional Imbalances . . . 118
Ringtail . . . 119
Splay Leg . . . 119
Trichobezoars . . . 119
Ulcerative Pododermatitis . . . 120
Vertebral Luxation or Fracture . . . 120
General Bibliography . . . 120

CHAPTER 6 • CASE REPORTS

Rabbits . . . 125
Guinea Pigs . . . 127
Hamsters . . . 129
Gerbils . . . 130
Mice . . . 131
Rats . . . 132
Suggested Answers . . . 133

Index . . . 143

Chapter 1

General Husbandry

Laboratory animal medicine is preventive medicine, and preventive medicine involves a conscientious, continuing concern for exemplary animal care. Diseases of laboratory animals, which can kill a pet, eliminate a colony, or ruin a valuable and long-term experiment, are relatively easy to prevent but difficult to cure.

The content of this chapter on general husbandry, the most important information in the entire text, will be referred to repeatedly in subsequent chapters on biology, syndromes, and diseases. Instill in the client a continuing regard for the principles of disease prevention, and discussions of syndromes, treatments, and specific diseases become superfluous.

Specific, detailed guidelines for establishing an acceptable level of animal care and for meeting the legal regulations of the United States Department of Agriculture (The Animal Welfare Acts of 1966 and 1970) and the recommended standards of the National Institutes of Health are contained in these publications:

Title 9: *Animal Products, Chapter 1, Subchapter A—Animal Welfare*, available upon request from the Deputy Administrator, Veterinary Services, Animal and Plant Health Service, United States Department of Agriculture, Hyattsville, MD 20782.

Guide for the Care and Use of Laboratory Animals (DHEW Publication No. NIH 74-23) available for 70 cents from the Superintendent of Documents, United States Government Printing Office, Washington, DC 20402.

Major Concerns in Husbandry

The major considerations in preventive management include general husbandry factors, circumstances predisposing to disease, and methods of facility sanitation.

HOUSING

Primary enclosures (cages, pens, and runs) should be structurally sound, in good repair, free of sharp or abrasive surfaces, easily cleaned, constructed to prevent escape and intrusion, and large enough to provide for normal postural adjustments, eating, and breeding. Housing areas should receive reliable electric power and potable water supply and should be uncluttered, easily sanitized, properly drained, well-lighted, well-ventilated, and heated. Transport vehicles should provide a housing environment comparable to the environment in stationary housing. Unpainted wood, which is difficult to sanitize, should not be used for housing.

DIET

Laboratory animals should be fed a wholesome, fresh, clean, nutritious, palatable diet on a regular basis and in adequate quantity. The feeding and watering devices should be clean, appropriate for the species and age of animal housed, and functional. Water should be fresh, clean, and always available in sipper-tube watering devices. Food should be stored in closed containers, kept at room temperature or less, observed for mold or vermin, and used while fresh.

PHYSICAL COMFORT

The provision of physical comfort extends to the animal care personnel as well as to the animals. Housed animals should be dry, clean, away from excessive noise, and maintained within an ambient temperature range of 18° to 29° C (65° to 85° F). Ideally the temperature should not vary more than 2 degrees from an average level, usually 22° C (72° F). The humidity should be maintained between 30% to 70% saturation and the light intensity between 100 and 125 foot candles. The room air changes, with fresh or filtered air, should not be fewer than 10 complete air changes per hour. Animals housed outdoors must be in well ventilated areas away from excessive drafts, dampness, direct sunlight, and temperature extremes. The cage area and height should permit normal postural adjustments, feeding, and reproductive behavior.

HEALTH

Facilities and caging should be physically cleaned and sanitized when necessary (usually 1 to 3 bedding changes per week). Vermin must be excluded from animal quarters. Different species and diseased animals should be housed separately, preferably in different rooms. Isolation and quarantine facilities must be available and utilized. Professional and technical personnel should regularly inspect animals for injury and disease. Stock and replacement animals should be obtained from reputable dealers.

IDENTIFICATION

The animals should be properly and clearly identified. Identification methods vary, but animals may be identified by cage cards, individual coat pattern, ear punch or notch (mice, rats, and hamsters), toe clip (mice), ear tag or stud (hamsters, guinea pigs, and rabbits), dye staining (on light colored fur), or tattooing (ear, tail, foot pad, or shaved flank). An ear notch-punch code is shown in Figure 1.

Factors Predisposing to Disease

Certain organic or environmental factors (often called "stressors") act to increase the exposure or reduce the resistance of the animal to disease agents. These factors must be considered in disease prevention efforts. Factors influencing disease susceptibility include environmental, host, experimental, and dietary variables.

ENVIRONMENTAL FACTORS

Climatic extremes
Climatic changes
Inadequate ventilation
High ammonia levels

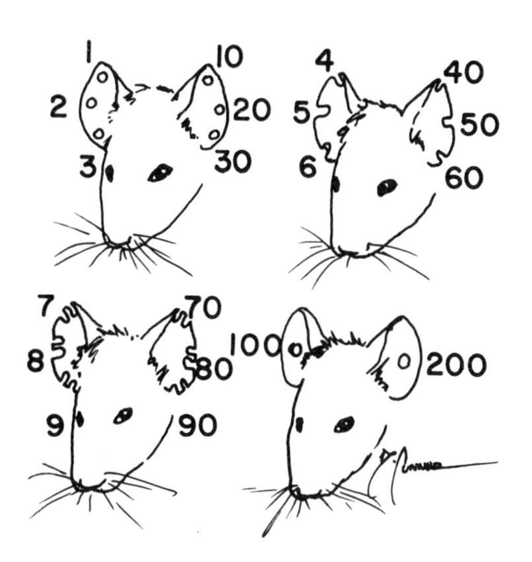

Fig. 1. Ear notch-punch code for the identification of rodents. These number codes are used in various combinations to produce the desired number.

Excessive drafts
Dampness
Personnel changes
Crowding
Improper bedding
Social hierarchies
Excessive noise
Improper lighting
Waste accumulation
Exposure to animal vectors

HOST FACTORS

Sex differences
Inherited mutations
Strain differences
Congenital abnormalities
Immune system deficiencies
Behavior and cycles
Immunologic status
Age
Obesity
Concurrent disease
Anorexia
Exercise lack
Lactation
Pregnancy
Nonspecific stressors

EXPERIMENTAL FACTORS

Restraint
Surgery
Drug effects
Neoplastic induction
Radiation effects
Pathogen inoculation
Bleeding

DIETARY FACTORS

Feed or water insufficient in quantity
- Insufficient amount supplied
- Feeders or waterers inaccessible, not recognized, not working, or cannot be operated by the age group or species involved
- Feed supplied in a form (hard pellets or soft mash) that cannot be eaten by age group or species involved
- Increased demand for nutrients because of pregnancy, lactation, heat, cold, diet composition, or disease
- Competition for feed
- Water frozen

Feed or water inadequate in quality
- Improper formulation
- Intended for another species
- Unpalatable
- Deteriorated
- Contaminated by insects, mold, bacteria, urine, feces

Dietary alterations
- Feed not recognized by the animal
- Intestinal flora alteration due to dietary change or antibiotic administration
- Change in gastrointestinal floral substrate at weaning

Facility Cleaning and Disinfection

Sanitation is a key operation in laboratory animal care. Clean cages are particularly important during pregnancy, lactation, and weaning, after the removal of sick animals, and preceding the introduction of new animals into the colony.

Use flowing, clean water, a commercial

detergent, and a brush or automatic washer to remove accumulated bedding, feed, urine, and feces from the cages, feeders, and waterers. Organic materials remaining on the cage will inhibit the antimicrobial activity of disinfectants.

After gross waste accumulations have been rinsed from the cage, wash with 82° C (180° F) water for at least 3 minutes or apply a disinfectant solution to all surfaces. Disinfectant solutions (phenolic disinfectants, quaternary ammonium compounds, and halogens) are available in farm supply or feed stores and from manufacturers. Instructions for use are on the labels of the bottled concentrates.

Disinfectants should be selected for broad spectrum activity, rapid kill effect, cleaning capacity, solubility, stability, residual activity, and lack of odor, toxicity, and irritability. Disinfectants selected should be effective in the presence of organic materials, soap, hard water, and varying pH levels and on porous, rough, or cracked surfaces.

Unfortunately, no single disinfectant meets all these criteria, and selection must be based on specific requirements. The effects of disinfectants vary with time of exposure, temperature and concentration of solution, and ionic content of the diluent. Important categories of microorganisms weakly or not affected by standard disinfectant solutions are the *Mycobacterium* spp, bacterial spores, and *Pseudomonas* spp. Viruses and fungi also vary in susceptibility to disinfectants.

Phenol derivative compounds, which are the disinfectants least affected by environmental agents, kill the vegetative forms of both Gram-positive and Gram-negative bacteria (excepting *Pseudomonas* spp, which require longer exposures and higher concentrations) in approximately 30 minutes. The germicidal activity is increased with the concentration and temperature of the solution. Phenolic compounds, emulsified at 1% to 5% in weakly acidic, soapy water, have some antifungal, sporicidal, and virucidal activity. Because of a residual odor and toxicity, phenolic derivatives are not used to disinfect feeders and waterers. Synthetic phenolic compounds, including pine oils, have greater activity and less odor than natural coal tar derivatives; but, as with other phenolic compounds, they are particularly toxic for cats.

Quaternary ammonium compounds are effective against Gram-positive bacteria but are considerably less effective in the presence of organic matter, soaps, and an acidic pH. These compounds are useful for general purpose disinfection and for cleaning feeders and waterers.

Halogen-bearing disinfectants, the hypochlorites and iodophors, are effective in acidic solutions, but they may stain or damage fabrics and have reduced activity in the presence of organic matter, soap, or detergent residues. Other disinfecting substances include 2% lye solution, formalin, ethylene oxide gas, and 10% ammonia solution.

Alkaline rabbit urine (pH 8.2) and guinea pig and hamster urine contain phosphate and carbonate crystals. When accumulated on the caging, these crystals form a scale that is difficult to remove. One of the available acidic products for removing the scale is Lime-A-Way (pH 2), manufactured by Economics Laboratory, Inc., Osborn Building, St. Paul, MN 55102.

Details for water acidification as a measure for reducing *Pseudomonas* contamination are discussed in the section on preventing disease in a mouse colony.

Formaldehyde Gas Fumigation

Formaldehyde gas fumigation, if preceded by thorough mechanical cleaning of the room and caging, is an effective method for eliminating parasites and vegetative bacterial forms. Spores, oocysts, and parasitic ova are better eliminated by mechanical cleaning and the application of a solution of formalin, ammonia, activated glutaraldehyde, or aldehyde-alcohol.

Before fumigation is attempted, the room must be free of animals, airtight, warmed to at least 21° C (70° F), and wetted to raise the relative humidity to 80% or more. Provisions should be made for exhausting the room without the entry of personnel.

Formaldehyde gas can be generated by:

1. Heating a small bucket of formalin on a hot plate. Provide for turning the hot plate on and off outside the area to be fumigated. Vaporize 1 ml formalin per 0.28 m^3 (1 ft^3). Seal the room for 8 hours.
2. Mixing, in a ceramic container, 400 ml formalin and 200 gm potassium permanganate for each 28.3 m^3 (1000 ft^3).
3. Heating Formaldegen (Vineland Poultry Labs., Vineland, NJ 08360) on a hot plate. Paraformaldehyde, unlike formalin, does not corrode metal. Instructions are on the package.

BIBLIOGRAPHY

American Association for Laboratory Animal Science: Disinfectants, Their Chemistry, Use, and Evaluation. Publication 74-5, Joliet, IL, AALAS, 1974.

Animals for Research. 9th Edition. Washington, DC, Institute of Laboratory Animal Resources, NAS-NRC, 1975.

Animal Models for Biomedical Research I, II, III, IV, V. Washington, DC, Institute of Laboratory Animal Resources, NAS-NRC, 1968-1974.

Bruce, H. M.: The water requirements of laboratory animals. J. Anim. Tech. Assoc., *1:*2-8, 1950.

Committee on Revision of the Guide for Laboratory Animal Facilities and Care: Guide for the Care and Use of Laboratory Animals. Washington, DC, Institute of Laboratory Animal Resources, NAS-NRC, 1972.

Festing, M., and Bleby, J.: A method for calculating the area and growing accommodation required for a given output of small laboratory animals. Lab. Anim., *2:*121-129, 1968.

Gay, W. I. (ed.): Methods of Animal Experimentation I, II, III, IV, V. New York, Academic Press, 1965-1974.

Hafez, E. S. E. (ed.): Reproduction and Breeding Techniques for Laboratory Animals. Philadelphia, Lea & Febiger, 1970.

ILAR News. Institute of Laboratory Animal Resources, NAS-NRC, 2101 Constitution Avenue, Washington, DC 20418.

Laboratory Animal Science. American Association for Laboratory Animal Medicine, 2317 West Jefferson, Suite 208, Joliet, IL, 60435.

Laboratory Animals. Journal of the Laboratory Animal Science Association, Laboratory Animals Ltd., 7 Warwick Court, London, WCIR 5DP.

Melby, E. C., and Altman, N. H. (eds.): Handbook of Laboratory Animal Science, 3 Vols. Cleveland, Chemical Rubber Company Press, 1974.

Mulder, J. B.: Animal behavior and electromagnetic energy waves. Lab. Anim. Sci., *21:*389-393, 1971.

Nutrient Requirements of Laboratory Animals. Publication 990. Washington, DC, NAS-NRC, 1972.

Nutrient Requirements of Rabbits. Publication 1194. Washington, DC, NAS-NRC, 1966.

Schoental, R.: Carcinogenicity of wood shavings. Lab. Anim., *7:*47-49, 1973.

Serrano, L. J.: Carbon dioxide and ammonia in mouse cages: Effect of cage covers, population, and activity. Lab. Anim. Sci., *21:*75-85, 1971.

Simmons, R. C.: Selected Topics in Laboratory Animal Medicine, Vol. II: The Design of Laboratory Animal Homes. Brooks Air Force Base, USAF School of Aerospace Medicine, Texas, 1973.

Short, D. J., and Woodnott, D. P. (eds.): The I.A.T. Manual of Laboratory Animal Practice and Techniques. Springfield, IL, Charles C Thomas, 1969.

Title 9: Animals and Animal Products, Chapter 1, Subchapter A—Animal Welfare. Hyattsville, U.S. Dept. of Agriculture, 1972.

Universities Federation for Animal Welfare: The UFAW Handbook on the Care and Management of Laboratory Animals. Baltimore, Williams & Wilkins, 1972.

Chapter 2

Biology and Husbandry

Chapter 2 covers selected topics in the biology and husbandry of rabbits, guinea pigs, hamsters, gerbils, mice, and rats. The topics are divided into nine categories: Taxonomy; Sources of Information; The Animal as a Pet; Housing; Feeding and Watering; Breeding; Disease Prevention; Public Health Concerns; and Unique Characteristics. A bibliography follows each species.

The Rabbit

The domestic rabbit, *Oryctolagus cuniculus,* which can be housed indoors or outdoors and fed a readily available pelleted chow, can serve as a pet, meat producer, or research animal. If rabbits are raised in clean wire cages, watered from sipper tubes, fed from hoppers, and protected from predators, they will lead long, productive, and healthy lives.

TAXONOMY

Taxonomic Position

Class:	Mammalia
Order:	Lagomorpha
Family:	Leporidae
Genus:	Oryctolagus
Species:	cuniculus

Oryctolagus cuniculus is the only genus of European rabbit. Hares (*Lepus*) and cottontails (*Sylvilagus*) are in different genera. Fertile, cross-genera matings do not occur. The European rabbit evolved in recent times in Southwestern Europe and was domesticated in the late Middle Ages.

Varieties of Rabbits Available

The American Rabbit Breeders Association (ARBA) lists 28 breeds and about 77 varieties of *Oryctolagus cuniculus.* Breeds vary in size, hair coat, and color and are selected for show, meat, fur, and research purposes. Representatives of the large breeds (5 kg or 10 lb and over) are the Flemish Giant and Checkered Giant. Among the medium-sized breeds (2 to 5 kg or 5 to 10 lb) are the Californian and New Zealand rabbits. The white (albino) New Zealand is the popular choice for meat, fur, and research production. Small breeds (under 2 kg or 5 lb) include the Dutch and Polish breeds, both popular as pets and often seen in research colonies.

SOURCES OF INFORMATION

Two booklets on rabbit husbandry (Agric. Handbooks 358 and 490) are available for purchase from the Superintendent of Documents, United States Government Printing Office, Washington, DC 20402.

A journal, *Countryside and Small Stock Journal,* is available from the editor, Jerome D. Belanger, Rt. 1, Box 239, Waterloo, WI 53594.

The American Rabbit Breeders Association, 1007 Morrissey Drive, Bloomington, IL 61701, has several publications on various aspects of rabbit raising.

The Ralston-Purina, Co., Checkerboard Square, St. Louis, MO 63188, and Carnation-Albers, 6400 Glenwood, Suite 300, Shawnee, KS 66202, publish information covering several aspects of rabbit husbandry and disease.

The standard reference text concerning the rabbit is *The Biology of the Laboratory Rabbit* by S. H. Weisbroth, R. E. Flatt, and A. L. Kraus. This text is published (1975) by Academic Press, Inc., 111 Fifth Avenue, New York, NY 10003.

Aeromedical Review 6-73, Selected Topics in Laboratory Animal Medicine, Volume XXI, *The Rabbit,* by R. J. Russell and P. W. Schilling is available from the National Technical Information Service, 5285 Port Royal Road, Springfield, VA 22161.

RABBITS AS PETS

Rabbits make excellent pets and can be house trained. Rabbits, especially older bucks, may bite, but biting is rare. They can inflict painful scratch wounds with their powerful rear limbs, but scratching usually occurs only if the rabbit is improperly restrained. Rabbit raising is a practical livestock project in suburban or urban areas, where permitted.

Life Span of Rabbit

A rabbit has a breeding life from 6 months to 3 to 4 years of age, an average life span of 6 to 7 years, and a possible longevity of 15 years.

Fig. 2. The two-hand grip for restraining a rabbit. Rabbits improperly restrained will struggle and may luxate or fracture the lumbar spine.

Commercial Market for Rabbits

Pet stores, local clients, research institutions, and a few meat processors purchase rabbits; however, the users' requirements for quality, quantity, delivery date, and price are often beyond the capacity of a small operation. One should not invest in a commercial rabbitry without establishing and understanding the actual and potential market.

Picking Up and Carrying a Rabbit

To carry a rabbit for a short distance, grasp the neck skin with one hand and support the rear quarters with the other hand (Fig. 2). For longer distances, the rabbit is placed on the bearer's forearm with its head concealed in the bend of the elbow (Fig. 3). If rabbits are improperly handled, they struggle and may break their backs or scratch the handler. If rabbits are carried for long distances, carrier boxes, with a door and handle, are recommended. Rabbit ears are sensitive and fragile; they must not be used to lift or restrain the animal.

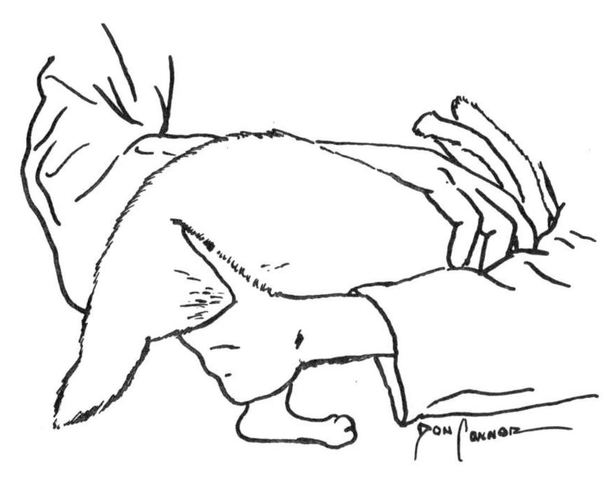

Fig. 3. Method of restraint for carrying a rabbit. The head and eyes are concealed in the elbow. Adapted by permission from Reproduction and Breeding Techiques for Laboratory Animals by E. S. E. Hafez (c) 1970 by Lea & Febiger, Philadelphia, Pennsylvania.

Fighting

Sexually maturing rabbits (over 3 mo) often attack one another. Mature bucks housed together will bite at the testes. As wounding and pseudopregnancies may result from group housing, mature rabbits should be separated and caged singly.

HOUSING

Plans for rabbit hutches are available from the representative feed companies mentioned previously. Complete wire cages with feeder and waterer attached are available from farm supply stores or mail order houses (Fig. 4). Laboratory caging is advertized in professional journals. Critical considerations in housing include structural strength, absence of sharp corners, size, wire bottom (14 gauge—2.5 × 1 cm or 1 × 0.5 in), ease of cleaning, protection from climatic extremes, portableness, and provision of self-feeding "J" hoppers and sipper-tube waterers. Adult rabbits (over 3 mo) are housed individually to prevent fighting, unwanted pregnancies, and pseudopregnancies.

Rabbits, like other animals, should be housed in areas commensurate with the animal's size and weight. Rabbits up to 2 kg (5 lb) should have 0.14 m^2 (1.5 ft^2) per animal; rabbits 2 to 5 kg (5 to 10 lb) require 0.28 m^2 (3 ft^2), and rabbits over 5 kg (10 lb) require 0.37 m^2 (4 ft^2) per animal. A doe with a litter requires an additional 0.19 m^2 (2 ft^2) per cage.

Rabbits are housed at temperatures between 6° and 29° C (40° to 85° F), with a recommended average at a constant level between 18° and 21° C (65° to 70° F).

Rabbits should not be housed near guinea pigs because of the possibility of transmission of *Bordetella bronchiseptica* to guinea pigs.

Outdoor Housing

Rabbits can be successfully housed outdoors if protected from cold (4° C or 40° F or less) by provision of an enclosed area within the cage. In hot weather (29° C or 85° F and above) cages should be cooled by adequate ventilation, shade, or overhead sprinklers. Rabbits, like guinea pigs, are susceptible to heat stroke. Rabbits must also be protected from predators, vermin, and excessive drafts.

Cleaning Rabbit Cages

Hanging wire cages remain relatively clean for weeks. Dropping pans should

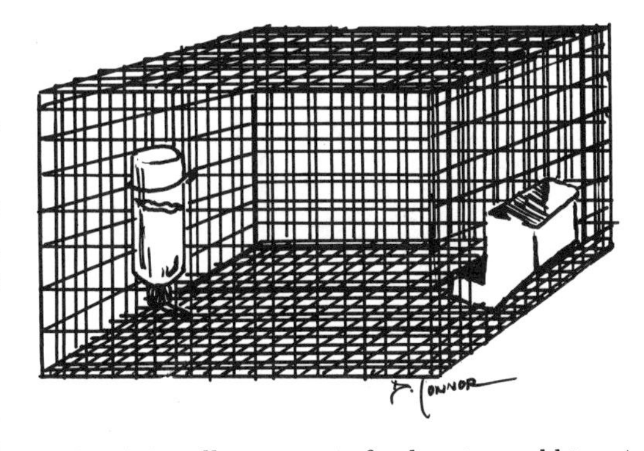

Fig. 4. An all-wire cage for housing rabbits. A water bottle and a feed hopper are attached to the cage.

be emptied daily, and enclosed hutches should be cleaned as needed or weekly. Cleaning is particularly important before parturition, during weaning, and after the removal of a sick animal. Alkaline rabbit urine contains a high concentration of crystals, which accumulate on cage surfaces and are difficult to remove. Detergents, disinfectants, lime-scale removers, and a stiff brush are used in cleaning cages. Lime-scale removers are acid solutions (vinegar, Lime-A-Way) that dissolve alkaline crystals.

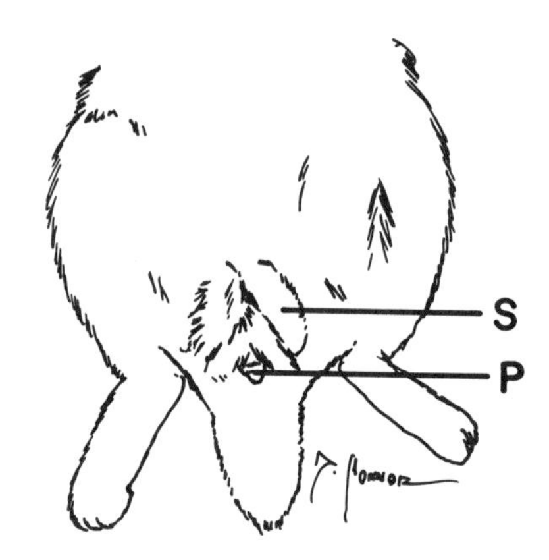

Fig. 6. External genitalia of the male rabbit. S = scrotal sac; P = tip of penis. The penis overlies the anus.

FEEDING AND WATERING

Rabbits should be fed a high quality, commercially prepared, pelleted, complete rabbit diet *ad libitum* unless obesity is a problem, in which case a high fiber (25%) or limited feed should be fed to nonbreeding bucks and does. A once daily feeding of an amount of feed that will maintain the desired body weight is recommended. High fiber diets and food restriction are not intended for pregnant or lactating does. Supplementation of a complete rabbit diet with salt, hay, table scraps, or antibiotics is not recommended unless indicated by a veterinarian or experienced husbandryman.

The amount of feed needed varies with the size of the rabbit. Rabbits will consume approximately 5 gm feed per 100 gm body weight per day (lactating rabbits may eat 2 to 3 times as much). With medium-sized breeds, up to 180 gm of feed (6 ounces or 1 cup pellets) may be adequate. Research rabbits may be maintained on 100 to 120 gm feed per day to prevent obesity. Rabbits should be supplied *ad libitum* with fresh, clean water (they drink approximately 10 ml per 100 gm body weight per day if nonpregnant and up to 90 ml per 100 gm if lactating).

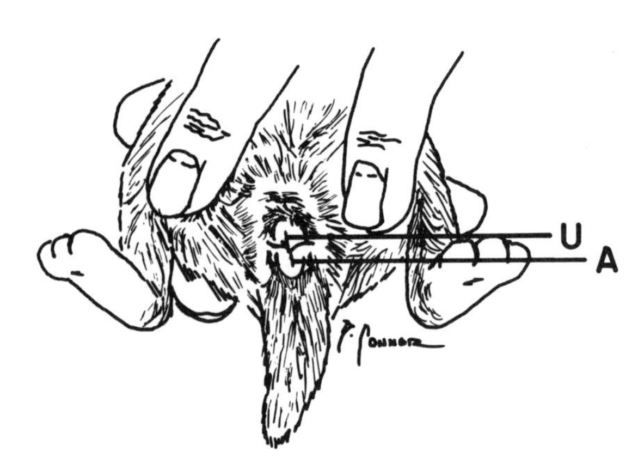

Fig. 5. External genitalia of a young female rabbit. U = urogenital orifice; A = anus. The perineal pouches lie laterally to the urogenital opening.

BREEDING

Sexing a Rabbit

Young males have a pointed, protruding genital opening, and the female has a short-slit opening (Fig. 5). The inguinal pouches are evident in the mature buck as elongated, hairless areas lateral and anterior to the penis (Fig. 6).

Breeding Program for Rabbits

The optimal breeding age for rabbits is between 6 months and 3 years. One buck is kept for every 10 does. Show rabbits are usually limited to three or fewer litters per year. In rabbitries where maximum production is important, does are rebred at 2 weeks postpartum and the young are weaned at 4 weeks, although later wearning, 6 to 8 weeks, promotes better growth and health. An intensive schedule, requiring good management, will produce up to eight litters a year, or approximately 256 lb of meat.

Small breeds (Dutch, Polish) are mated at 5 months; medium breeds (New Zealand White, Californian) at 7 months, and large breeds (Flemish, Checkered) at 9 months.

Receptivity for Mating

The doe, although an induced ovulator with no estrous cycle, will exhibit a short (1 to 2 day) loss of receptivity every 4 to 17 days. Receptivity is tested by carrying the doe to the buck and leaving them together for 15 minutes. If fighting or no mating occurs, try another buck or pair the rabbits the next day. Congestion of the vulva is not a reliable indicator of receptivity.

Estrous Cycle in Rabbits

Rabbits do not have an estrous cycle. They do exhibit a short (1 to 2 day) lack of receptivity every 4 to 17 days. This short "rest" period is due to the periodic atresia of mature follicles.

Gestation Period in Rabbits

The rabbit's gestation period lasts 28 to 35 days, with an average of 32 days.

Determination of Pregnancy

Gentle palpation of the ventral abdomen at 10 to 14 days gestation will reveal marble-sized swellings in the uterus (Fig. 7). The hand is placed between the hind legs, just in front of the pelvis. With thumb on one side of the abdomen and fingers on the other, a light pressure is exerted as the hand is gently moved back and forth. Rabbits pregnant 14 to 21 days will usually refuse a buck. Pregnant rabbits begin hair pulling and nest building within the last 3 days of gestation.

Pseudopregnancy

Pseudopregnancy is common in rabbits following sterile matings, excitement by nearby bucks, or mounting by other does. Ovulation is followed by a persistent corpus luteum, which lasts 15 to 17 days and

Fig. 7. Palpation of a rabbit for pregnancy. Rabbits are palpated for pregnancy between 10 and 14 days of gestation.

secretes the progesterone that causes mammary gland enlargement and nest-building behavior.

Nest Building

During the last week of pregnancy, a clean nest box with a solid floor is placed in the cage. Straw or bedding in the box provides a base for the hair nest built by the rabbit. Nests may be split or have too little or too much hair.

Size of Litter

Rabbit litter size varies from 4 to 10 with an average of 7 to 8. Smaller breeds or older rabbits have fewer young.

Abandonment and Cannibalism

Cannibalism is rare in rabbits. When it occurs, it is associated with dead or deformed young or with a hyperexcitable, primiparous doe. Young rabbits may be abandoned if the nest is split, the young crawl out of the nest, the doe has agalactia or mastitis, or the doe is excessively disturbed.

Hand Rearing

Orphaned rabbits under 3 weeks can be fed warmed (29° C or 85° F) orphaned puppy formula by syringe, doll bottle, or gastric intubation (No. 5 French catheter). The liquid is given slowly three times a day. Give up to 5 ml per day the first week, 15 ml per day the second week, and 25 ml per day the third week. Aspiration pneumonia and diarrhea often result from hand feeding. The young should be kept in a warm area and massaged daily with a wetted, warm cotton swab to facilitate defecation and urination. The young may also be fostered to a lactating mother if the orphans are under 2 weeks of age and within 2 days of the host litter. Mortality of the fostered litter may be high (over 40%). Rabbit milk contains approximately 15% protein, 2.5% mineral, 10% fat, and 2% sugar.

DISEASE PREVENTION

Observe the recommendations for general husbandry and disease prevention listed in Chapter 1. Common management procedures with rabbits include periodic trimming of teeth and toenails and the exclusion of dogs and other animal pests from the rabbitry. A common problem with rabbits, and other laboratory animals raised on wire, is the catching and injuring of the limbs in the cage floor. Rabbits trapped in this way may sit for hours or days with few signs of distress.

PUBLIC HEALTH CONCERNS

Diseases of major public health importance in domestic rabbits are rare. Such diseases include salmonellosis, tularemia, tuberculosis, and toxoplasmosis. Rabies vaccinations (killed vaccine) are not given except in cases of high exposure risk.

ANATOMIC AND PHYSIOLOGIC CHARACTERISTICS

Coprophagy. Rabbits ingest soft feces of cecal origin passed via the anus. As these feces are ingested directly from the anus, placing the rabbits on wire does not impede coprophagy. This process is normal and contributes to the dietary efficiency of the rabbit. Coprophagy must be considered in nutritional or toxicologic experiments.

Induced Ovulation. Ovulation follows copulation by approximately 10 hours. Ovulation may be induced with human chorionic gonadotropin.

Atropinesterase Enzyme Activity. The presence of a serum atropinesterase in 30% or more of domestic rabbits affects the response to administration of atropine. The enzyme hydrolyzes atropine.

Light Skeleton. A rabbit's skeleton comprises 7% of its body weight compared to 13% for a cat. Fractures are common, particularly of the rear limb bones and the lumbar spine. They are due in part to the greater muscle mass and weight of the animal in domestic compared to wild conditions.

Turbid Urine. Rabbit urine is alkaline, because of the plant diet, and contains carbonate and phosphate crystals. The cloudy yellow to red-brown urine, which may be mistaken for a purulent discharge, causes a scale on caging.

Wide Immune Competency. Rabbits produce high levels of circulating antibodies to a wide variety of antigens.

Pseudoeosinophils. Rabbit neutrophils have intracytoplasmic, irregularly sized, eosinophilic granules and are easily confused with eosinophils in blood films and tissue sections.

Large Eyes, Ova, and Ear Veins. These structures, relatively for the eyes, absolutely for ova, are among the largest in mammals. The ear veins make intravenous injections relatively easy. The rabbit heart, on the other hand, is small in proportion to total body weight.

Four Upper and Two Lower Incisors. The two rear upper incisors are a distinguishing anatomic feature of the Lagomorpha compared to rodents.

Divided Cervix. The rabbit cervix and uterus are divided; the cervix has two internal and external ora.

Arteriosclerosis. Spontaneous arteriosclerosis occurs in rabbits and has a probably genetic-dietary etiology. Since vitamin D toxicity has been suggested as a possible etiology for some forms of the disease, guinea pig chow and processed cow's milk, both vitamin D fortified, are contraindicated for rabbits.

Pyrogen Testing. Rabbits are commonly used for determining pyrogenic effects of organisms or chemicals.

BIBLIOGRAPHY

Barone, R., et al.: Atlas of Rabbit Anatomy. Paris, Masson, 1973.

Broadfoot, J.: Handrearing rabbits. J. Inst. Anim. Tech., *20*: 91-99, 1969.

Chapin, R. E., and Smith, S. E.: Calcium requirement of growing rabbits. J. Anim. Sci., *26*:67-71, 1967.

Cheeke, P. R.: Feed preferences of adult male Dutch rabbits. Lab. Anim. Sci., *24*:601-604, 1974.

Denenberg, W. H., Zarrow, M. X., and Ross, C.: The behavior of rabbits. *In* The Behavior of Domestic Animals. Edited by E.S.E. Hafez. Baltimore, Williams and Wilkins, 1969.

Fox, R. R., and Crary, D. D.: A simple technique for the sexing of newborn rabbits. Lab. Anim. Sci., *22*:556-558, 1972.

Henschel, M. J., and Coates, M. E.: The toxicity of cow's milk to infant rabbits. Proc. Nutr. Soc., *33*:112A, 1974.

Herndon, J. F., and Hove, E. L.: Surgical removal of the cecum and its effect on digestion and growth in rabbits. J. Nutr., *57*:261-270, 1955.

Hirzel, R.: Note on the effect of condition on the colour of body fat. J. Agric. Sci., *25*:541-544, 1935.

Menzies, W., and Moss, A.: The mating of domestic rabbits. J. Anim. Tech. Assoc., *11*:7-9, 1960.

Mroueh, A., and Mastroianni, L.: Insemination via the intraperitoneal route in rabbits. Fertil. Steril., *17*:76-82, 1966.

Russell, R. J., and Schilling, P. W.: Selected Topics in Laboratory Animal Medicine, Vol. XXI: The Rabbit. Brooks Air Force Base, Texas, USAF School of Aerospace Medicine, 1973.

Sawin, P. B., and Glick, D.: Atropinesterase, a genetically determined enzyme in the rabbit. Proc. Natl. Acad. Sci., *29*:55-59, 1943.

Thacker, E. J., and Brandt, C. S.: Coprophagy in the rabbit. J. Nutr., *55*:375-386, 1955.

Weisbroth, S. H., Flatt, R. E., and Kraus, A. L.: The Biology of the Laboratory Rabbit. New York, Academic Press, 1974.

Weisbroth, S. H., and Scher, S.: The establishment of a specific pathogen free rabbit breeding colony. Lab. Anim. Care, *19*:795-799, 1969.

Westmann, D. S., and Pleasants, J. R.: Rearing of germ-free rabbits. Proc. Anim. Care Panel, 9:47-54, 1959.

The Guinea Pig

The guinea pig, *Cavia porcellus*, is a gentle rodent often encountered as a pet or research animal. Guinea pigs are remarkable for their fastidious eating habits, dependence on exogenous vitamin C, and long gestation period.

TAXONOMY

Taxonomic Position

Class:	Mammalia
Order:	Rodentia
Family:	Cavidae

Genus: Cavia
Species: porcellus

Cavia porcellus, the guinea pig or cavy, evolved in South America and is a member of the suborder Hystricomorpha (chinchilla, porcupine). Other rodent suborders include the Sciuromorpha (squirrels) and Myomorpha (rats, mice, hamster, gerbils).

Varieties of Guinea Pigs Available

The most common pet and laboratory variety is the English or short-haired guinea pig. The Duncan-Hartley and Hartley strains are representative lines of the English variety. Inbred strain 2 and 13 guinea pigs are popular research animals. Other varieties are the Abyssinian, with short, rough hair arranged in whorls or rosettes, and the Peruvian long hair or "rag mop" variety. Guinea pigs may be monocolored, bicolored, or tricolored. The varieties interbreed, and a profusion of colors and hair lengths is possible.

SOURCES OF INFORMATION

The booklet *Raising Guinea Pigs* (Leaflet No. 466-1960) is available for 50 cents from the Superintendent of Documents, U.S. Government Printing Office, Washington, DC 20402.

An illustrated booklet *Enjoy your Guinea Pig* is available from pet stores or from the Pet Library, Ltd., 600 South 4th Street, Harrison, NJ 07029.

The Guinea Pig Newsletter is distributed upon request by the Medical Research Council, Laboratory Animal Centre, Woodmansterne Road, Carshalton, Surrey, U.K. SM5 4EF.

The National Technical Information Service, 5285 Port Royal Road, Springfield, VA 22161, distributes a booklet in their Selected Topics in Laboratory Animal Medicine series entitled *The Guinea Pig* (Vol. XXII).

The Anatomy of the Guinea Pig, by Gale Cooper and Alan L. Schiller, was published (1975) by the Harvard University Press, 79 Garden Street, Cambridge, MA 02138.

The standard reference text concerning the guinea pig is *The Biology of the Guinea Pig* edited by J. E. Wagner and P. J. Manning. This text is published (1976) by Academic Press, Inc., 111 Fifth Avenue, New York, NY 10003.

GUINEA PIGS AS PETS

Guinea pigs, if gently handled, make excellent pets. They rarely bite or scratch but do scatter feed and bedding. Guinea pigs seldom climb or jump out of open pens. They respond favorably to frequent handling and become conditioned to squeal before reward situations.

Life Span of Guinea Pig

The breeding life of a laboratory housed guinea pig is from 18 months to 4 years, and they have been known to live 9 years; however, they rarely survive in the home more than 3 years and their litter size is reduced to 1 or 2 per litter by 2 years of age.

Commercial Market for Guinea Pigs

Sales outlets for guinea pigs include pet stores, fanciers, and research institutions. These markets often have specific requirements for quality, quantity, age, sex, price, health status, and delivery date. The small producer often finds these requirements difficult to meet. As with other businesses, the market should be determined before a large investment is made in animals, facilities, and registration with the United States Department of Agriculture.

Picking Up and Carrying a Guinea Pig

Guinea pigs are lifted by grasping the trunk with one hand while supporting the rear quarters with the other hand (Fig. 8). Support is particularly important with adult and pregnant animals. The "grasp-

Fig. 8. Restraint of a guinea pig. The hand placed under the body supports the animal without causing injury to the thoracic viscera.

ing" hand should be under the chest and abdomen and the "supporting" hand under the rear feet or the hindquarters (Fig. 9). Injured lungs may result from grabbing the animal over the back.

Fighting

Although colony or harem breeding arrangements are often successful, strange males placed together, especially in crowded conditions or in the presence of a female, will fight. Older animals frequently chew on the ears and hair of young or subordinate animals.

HOUSING

Guinea pigs may be housed in colony pens on the floor, in tiered bins, or in

Fig. 9. Restraint of a pregnant guinea pig. The hand beneath the rear quarters prevents struggling and supports the heavy body.

large "shoe box" cages. Young and breeder animals will occasionally climb out of the bins. Cages may be plastic, metal, or wire, but if guinea pigs not raised on wire are placed on wire, limbs are often broken. Adult animals may be placed on "rat mesh" (0.85 cm sq or 2 wires/in). Some guinea pig breeders prefer the square mesh, but others recommend a rectangular 1.3 × 3.8 cm (0.5 × 1.5 in) mesh. Open bins should have sides 25 cm (10 in) high, and covered cages should have sides 17.5 cm (7 in) high. Breeding animals should be provided with at least 652 cm^2 (101 in^2) of floor space per animal.

Beddings of wood shavings, shredded paper, or other plant origin material may be used; however, wood shavings and circumanal gland secretions can form a hard wad of material which irritates the prepuce.

Guinea pigs are housed at temperatures between 18° and 29° C (65° to 85° F) with the median ambient temperature at 24° ± 1° C (75° ± 2° F). High ambient temperatures without adequate air flow predispose to heat stress; low temperatures and wet bedding predispose to pneumonia. The environmental humidity should be between 30% and 70% saturation.

Outdoor Housing

If the animals are raised according to the regulations of the Animal Welfare Acts, guinea pigs can not be housed outdoors without the specific permission of the Director of the Animal Health Division, Agricultural Research Service, Hyattsville, MD 20782. At temperatures below 18° C or above 29° C, susceptibility to disease increases and production decreases. The young, particularly those on wire, are susceptile to cold stress.

Cleaning Guinea Pig Caging

Guinea pigs are messy housekeepers. Cages, pens, and feeding receptacles should be cleaned and sanitized at least once a week. Satisfactory cleaning procedures include washing the cages with a detergent and hot water (82° C or 180° F) or with a disinfectant solution followed by a thorough rinse.

FEEDING AND WATERING

Guinea pigs are notorious at chewing on and otherwise blocking sipper-tube waterers. They mix dry feed and water in their mouths and pass the slurry into the sipper tube, thereby blocking the tube. Guinea pigs suck from sipper tubes while rats lap; if the large, open-ended rat tube is used with guinea pigs, water will accumulate in the cage. As guinea pigs may defecate into their feed and water crocks, feeders and waterers should be suspended above the bedding.

Guinea pigs, which are fastidious eaters and may refuse to eat or drink if the consistency or taste of the feed or water or the feeding or watering devices are changed, should receive a freshly milled and properly stored pelleted, complete guinea pig chow. Whether this feed should be supplemented with hay or fresh greens is a matter of debate. Adult guinea pigs will consume approximately 5 gm feed and 10 to 40 ml water per 100 gm body weight daily; these amounts, however, vary with ambient temperature, breeding status, food and water wastage, and humidity. Guinea pig chows contain approximately 18% crude protein and 16% crude fiber.

Guinea pigs require dietary ascorbic acid at approximately 10 mg/kg body weight per day for maintenance and 30 mg/kg body weight per day for pregnancy. If ascorbic acid is not supplied in the feed, vitamin C may be added to the water (200 mg/L) or each guinea pig may be fed approximately 50 gm (one handful) of fresh cabbage daily. In water in an open crock the activity of vitamin C in

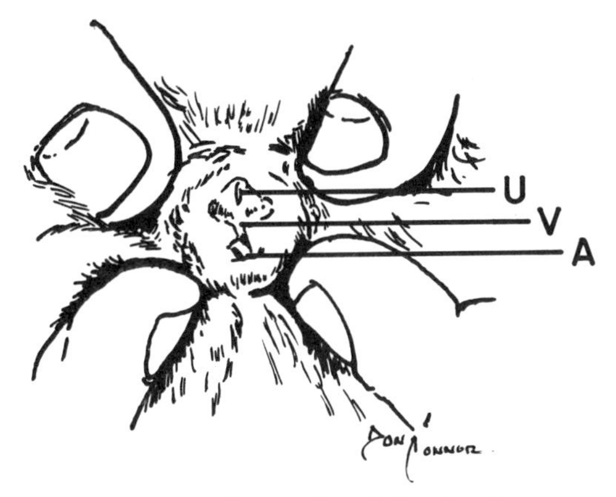

Fig. 10. External genitalia of the female guinea pig. U = urethral orifice; V= vaginal closure membrane; A = anus.

solution decreases as much as 50% in a 24-hour period.

BREEDING

Sexing a Guinea Pig

The male guinea pig has no genital opening between the urethral orifice and the anus. The female has a shallow, U-shaped break (the vaginal membrane) in this space. To reveal the vaginal membrane, place the thumb and forefinger of each hand on either side of the genital ridge. The fingers and thumbs are drawn apart gently to reveal the vaginal membrane (Fig. 10). In males the penis and testes may be palpated or the penis extruded by gentle digital pressure (Fig. 11).

Breeding Program for Guinea Pigs

One boar is housed in an adequate-sized pen (652 cm^2 or 101 in^2 per animal) with 3 to 10 sows. In intensive breeding systems, which utilize the postpartum estrus, the sows and boars remain together, and the young are removed at weaning. In a nonintensive system the sows are removed for parturition and rebred after weaning. If the young are born on wire, they should be removed to a finer mesh to prevent foot injury. Young guinea pigs are weaned at 1 to 3 weeks (150 to 250 gm).

Males are used as breeders at 550 gm (4 to 5 mo) and females at 500 gm (3 to 5 mo). Sows bred for the first time after approximately 9 months may experience dystocias due to the fusion and ossification of the pubic symphysis.

Receptivity for Mating

Boars and sows are paired at approximately 550 gm and 500 gm respectively. Since the pair is housed together, detection of estrus is not necessary unless timed matings are desired. In estrus lordosis and vaginal membrane rupture occur. If timed pregnancies are desired, mating can be detected by observation of the vaginal plug or sperm in a vaginal smear.

Estrous Cycle in Guinea Pigs

The estrous cycle in the guinea pig lasts 14 to 19 days. Sows have a postpar-

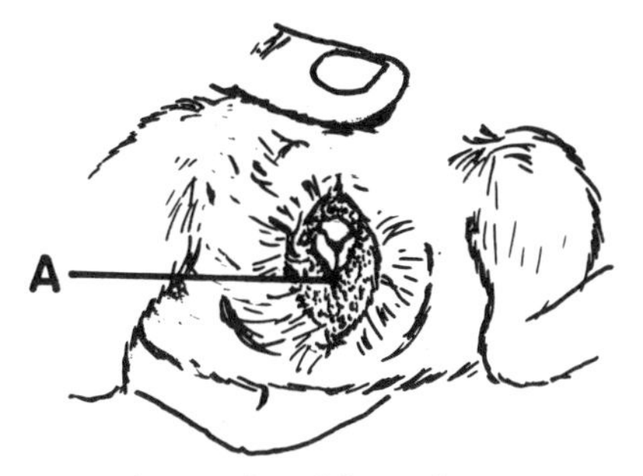

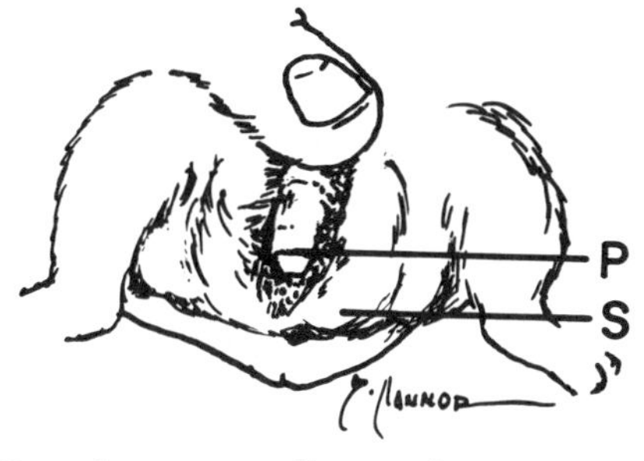

Fig. 11. External genitalia of the male guinea pig. Digital pressure will cause the extrusion of the penis. A = anus; P = penis; S = scrotal sac.

tum estrus at which time 60% to 80% will become pregnant if mated.

Gestation Period in Guinea Pigs

The gestation period in guinea pigs lasts from 59 to 72 days with an average between 63 and 68 days. The length of gestation depends upon the number in the litter: 2 = 70 days; 3 = 68 days; 4 = 66 days; 5 = 64 days.

Determination of Pregnancy

Fetuses may be palpated from 4 to 5 weeks to term. If the pre-pregnancy weight is known, the sow will nearly double her weight during the gestation period. Within the last week of pregnancy the pubic symphysis gradually separates to nearly 2 cm.

Pseudopregnancy

Pseudopregnancy is rare in guinea pigs. When it occurs, it lasts approximately 17 days.

Nest Building

No nests are built. Guinea pigs are precocious at birth and are delivered onto the cage floor.

Size of Litter

Guinea pigs may deliver from 1 to 6 young, but 3 to 4 is an average litter. Litters of 5 or more are usually born dead.

Abandonment and Cannibalism

The young are precocious (haired, eyes open, teethed) and can begin eating solid food within a few hours of birth. Cannibalism of healthy young is rare. The young, however, may be trampled if the adults are overcrowded or become excited. If the sow abandons the young, the young may be able to eat but viability is decreased. The problem with trampling is eliminated if the young and sow are removed to a nursery before or soon after birth.

Hand Rearing

As guinea pig young will begin eating solid food during the first few days postpartum, hand rearing is not difficult. During the first few days the neonates may be fed guinea pig chow softened with water or cow's milk. If given the opportunity, the young will nurse mothers other than their own.

DISEASE PREVENTION

Observe the general husbandry and disease prevention admonitions listed in Chapter 1. Guinea pigs scatter their bedding into their feed and water crocks, are susceptible to *Bordetella* pneumonia, have an absolute requirement for exogenous ascorbic acid, and are fastidious eaters. If the food is changed, guinea pigs may refuse to eat or drink. Because rabbits often have asymptomatic *Bordetella* infections, guinea pigs and rabbits should not be housed in the same room.

PUBLIC HEALTH CONCERNS

Diseases of major public health significance in guinea pigs are rare. Guinea pigs may harbor bacteria (*Bordetella, Salmonella, Pseudotuberculosis, Streptococcus, Diplococcus*) which are potential but unusual human pathogens. Rabies vaccinations in guinea pigs are not indicated except under conditions of increased risk of exposure.

ANATOMIC AND PHYSIOLOGIC CHARACTERISTICS

Exogenous Vitamin C Requirement. This characteristic, shared with primates, provides a model system for collagen synthesis.

Long Gestation Period. Germ-free, cesarean derivation is relatively easy to accomplish with the late-term guinea pig fetus. Neonates are precocious, nearly self-sufficient, and require little hand rearing if removed from the dam at birth. Endocrine control of gestation in the guinea pig is similar to that of the horse, monkey, and man in that the gestation period may be divided into trimesters of about 3 weeks' duration. The guinea pig is therefore an animal of choice for studying the effects of hormones and endocrine glands on pregnancy.

Histamine Sensitivity. Histamine interacts with the bronchiolar smooth muscle to cause rapid, lethal contraction.

Kurloff Bodies. The Kurloff body is an intracytoplasmic inclusion body occurring in certain monomorphonuclear white blood cells of the guinea pig. These bodies are more common in the female and increase with a rise in estrogen levels, which occurs in late pregnancy.

Foreign Protein Sensitization. Guinea pigs produce antibodies to specific proteins, and production of anaphylaxis is an indicator of the presence or absence of small amounts of antigen. Unlike the rabbit or chicken, injected antibodies protect a guinea pig from anaphylaxis.

High Level of Complement. Guinea pigs are a good source of complement for serologic investigators.

BIBLIOGRAPHY

Bacsich, P., and Wyburn, G. M.: Observations on the estrus cycle of the guinea pig. Proc. R. Soc. Edinb. (Biol.), *60*:33-39, 1939.

Bauer, F. S.: Glucose preference in the guinea pig. Physiol. Behav., *6*:75-76, 1971.

Beauchamp, G., Jacobs, W. W., and Hess, E. H.: Male sexual behavior in a colony of domestic guinea pigs. Am. Zool., *11*:618, 1971.

Carter, C. S.: Effects of olfactory experience on the behavior of the guinea pig (*Cavia porcellus*). Anim. Behav., *20*:54-60, 1972.

Christensen, H. E., Wanstrup, J., and Ranløv, R.: The cytology of the Foa-Kurloff reticular cells of the guinea pig. Acta Pathol. Microbiol. Scand. Suppl., *212*:15-24, 1970.

Dean, D. J., and Duell, C.: Diets without green food for guinea pigs. Lab. Anim. Care, *13*:191-196, 1963.

Gerall, H. D.: Effect of social isolation and physical confinement on motor and sexual behavior of guinea pigs. J. Pers. Soc. Psychol., *2*:460-464, 1965.

Goy, R. W., Hoar, R. M., and Young, W. C.: Length of gestation in the guinea pig with data on the frequency and time of abortion and still birth. Anat. Rec., *128*:747-757, 1957.

Jackway, J. S.: Inheritance of patterns of mating behavior in the male guinea pig. Anim. Behav., *7*:150-162, 1959.

Jacobs, W. W.: Male-female associations in the domestic guinea pig. Anim. Learning Behav., *4*:77-83, 1976.

King, J. A.: Social relations of the domestic guinea pigs living under seminatural conditions. Ecology, *37*:221-228, 1956.

McKeown, T., and Macmahon, B.: The influence of litter size and litter order on length of gestation and early postnatal growth in guinea pigs. Endocrinology, *13*:195-200, 1956.

Obeck, D. K.: Selected Topics in Laboratory Animal Medicine, Vol. XXII: The Guinea Pig. Brooks Air Force Base, Texas, USAF School of Aerospace Medicine, 1974.

Rowlands, I. W.: Postpartum breeding in the guinea pig. J. Hyg., *47*:281-287, 1949.

Stern, J.: Litter size and weight gain of neonatal guinea pigs. Psychol. Rep., *28*:981-982, 1971.

Townsend, G. H.: The guinea pig: general husbandry and nutrition. Vet. Rec., *96*:451-454, 1975.

Wagner, J. C., and Manning, P. J. (eds.): The Biology of the Guinea Pig. New York, Academic Press, 1976.

The Hamster

The golden hamster, *Mesocricetus auratus,* is popular as a pet and research animal. Hamsters are known for their short gestation period, pugnacious dispositions, cheek pouches, large immature litters, and ability to escape confinement. Hamsters do well on pelleted rodent chows and usually remain healthy and active over their short lifetime.

TAXONOMY

Taxonomic Position of the Hamster

Class:	Mammalia
Order:	Rodentia
Family:	Cricetidae
Genus:	Mesocricetus
Species:	auratus

Mesocricetus auratus, the golden or Syrian hamster, which is found wild in the Middle East, is the hamster most often seen as a pet or research animal. The Chinese (dwarf) hamster (*Cricetulus griseus*) and the European hamster (*Cricetus cricetus*) are infrequently seen in research colonies and almost never as pets.

Varieties of Golden Hamsters Available

Varieties include agouti, cinnamon, cream, white, piebald, and the long-haired "teddy bear."

SOURCES OF INFORMATION

A booklet *Hamster Raising* (Leaflet No. 250) can be obtained for 5 cents from the Superintendent of Documents, United States Government Printing Office, Washington, DC 20402. This leaflet was last printed in 1960.

Two booklets, *Know Your Hamster* and *Enjoy Your Hamster*, are available from The Pet Library, Ltd., 600 South 4th Street, Harrison, NJ 07029.

The standard reference text on the golden hamster is *The Golden Hamster* (1968) by R. A. Hoffman, P. F. Robinson, and H. Magalhaes and published by the Iowa State University Press, Press Building, Ames, IA 50010.

The Aeromedical Review (5-75) Volume XXIV, *The Hamster*, is available from the National Technical Information Service, 5285 Port Royal Road, Springfield, VA 22161.

Fig. 13. Scruff-of-the-neck grip for picking up and restraining a hamster. Because of the cheek pouches, a hamster has ample loose skin about the neck.

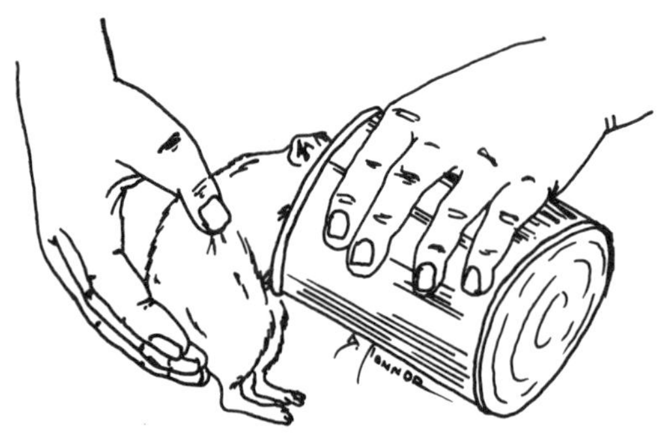

Fig. 12. Picking up and restraining a hamster with a small container. Adapted by permission from The Golden Hamster by R. A. Hoffman, P. F. Robinson, and H. Magalhaes, (c) 1968 by the Iowa State University Press, Ames, Iowa.

HAMSTERS AS PETS

If hamsters are handled gently and often, they become good pets; however, if roughly handled or startled, they may bite. Hamsters are adept at chewing on and escaping from their cages.

Life Span of Hamster

Female hamsters older than 14 months produce progressively smaller litters. Hamsters usually have a life span of 18 to 24 months, but older individuals have been reported.

Commercial Market for Hamsters

Hamsters are purchased by pet shops and research institutions; but, as with other laboratory animals, consumers often have specific requirements for quantity, health, delivery date, age, and sex. Before a large investment is made in animals and facilities, the market should first be understood and determined.

Picking Up and Carrying a Hamster

Hamsters often bite if roughly handled or awakened. Hamsters may be picked up in a small can (Fig. 12), by the loose skin of the neck (Fig. 13), by cupping the animal in open hands (Fig. 14), or by placing fingers over the back, sides, and tail (Fig. 15). A leather glove is used if the hamsters are not accustomed to being handled or if the handler is afraid of the animals.

Fighting

Except for the few hours of estrus occurring once during the 4-day cycle, the female will usually attack the newly introduced male. Breeder male hamsters are frequently injured by unreceptive females. Following copulation, the male is removed from the breeding cage. Females may fight other females, and males other males. Hamsters fight less often if housed together before sexual maturity or awakened simultaneously from anesthesia in a neutral cage.

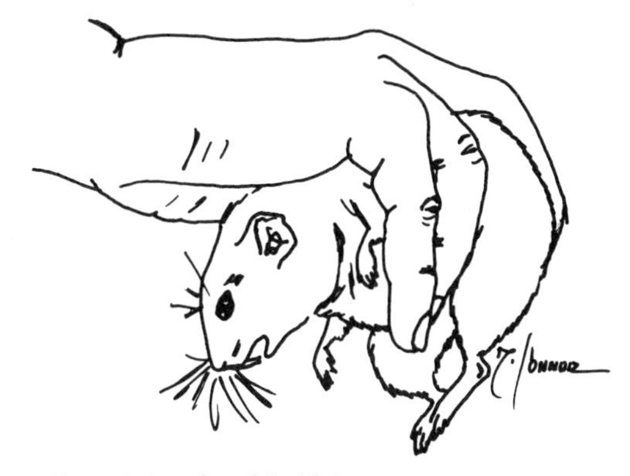

Fig. 15. One-hand hold for restraining a hamster. The thumb and third finger grasp the body. Adapted by permission from The Golden Hamster by R. A. Hoffman, P. F. Robinson, and H. Magalhaes, (c) 1968 by the Iowa State University Press, Ames, Iowa.

Fig. 14. Two-hand technique for picking up and restraining a hamster. Adapted by permission from The Golden Hamster by R. A. Hoffman, P. F. Robinson, and H. Magalhaes, (c) 1968 by the Iowa State University Press, Ames, Iowa.

HOUSING

Several types of hamster cages are available, some equipped with a variety of wheels, tunnels, and nesting areas. In research colonies, hamsters are usually housed in plastic, solid-bottom "shoebox" cages (Fig. 16), fed from a wire hopper, and watered from sipper-tube bottles or automatic waterers. Because hamsters have blunt noses, they may have difficulty eating from the slotted, sheet metal hoppers used for mice and rats. Beddings of wood shavings or other plant products are used. Hamsters chew plastic, wood, and soft metals and will readily escape from poorly secured or constructed cages.

An adult hamster (over 100 gm) re-

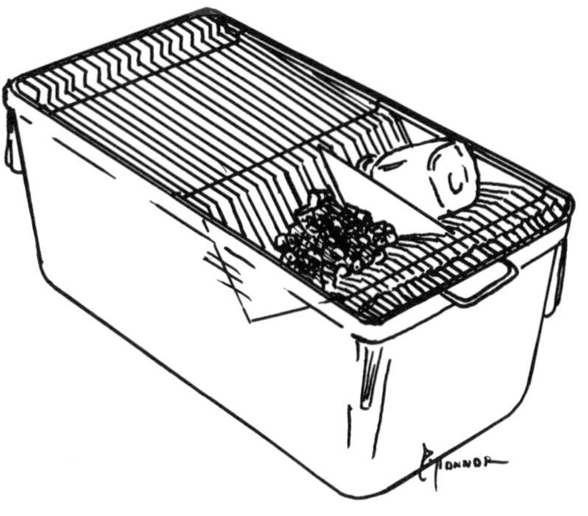

Fig. 16. The familiar "shoe-box" style of cage used for housing laboratory rodents.

quires a floor area of at least 123 cm² (19 in²) and a cage height of at least 15 cm (6 in). A female with litter requires approximately 786 cm² (121 in²) of floor space. Hamsters are housed at temperatures between 18° and 29° C (65° to 85° F) with a suggested level at 22° ± 1° C (72° ± 2° F). The environmental humidity should be between 30% and 70% saturation.

Outdoor Housing

The Animal Welfare Acts do not permit outdoor housing for hamsters. Small, domesticated rodents are stressed by climatic extremes and are adept at escaping from cages or enclosures.

Cleaning Hamster Cages

Hamster cages are washed when dirty or once or twice weekly with detergent and hot (82° C or 180° F) water or with a nontoxic, effective disinfectant and thorough rinse. Bottles and hoppers are cleaned at the same time as the cages.

FEEDING AND WATERING

Hamsters should be fed a quality, commercially prepared, complete pelleted rodent chow *ad libitum*. Fresh, clean water must be continually supplied. Young begin drinking water at 10 days of age, and the sipper tube should be accessible to the small animals. Hamsters eat approximately 10 gm feed per 100 gm body weight and drink approximately 14 ml water per 100 gm body weight. Laboratory rodent feed contains around 24% protein and 5% fiber.

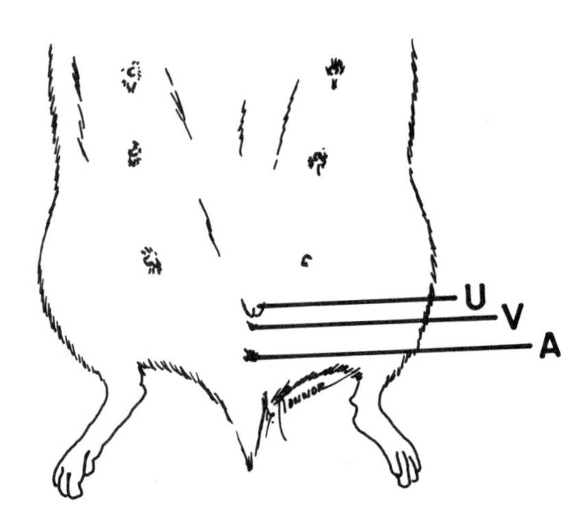

Fig. 18. External genitalia of the female hamster. U = urethral orifice; V = vaginal orifice; A = anus. Adapted by permission from The Golden Hamster by R. A. Hoffman, P. F. Robinson, and H. Magalhaes, (c) 1968 by the Iowa State University Press, Ames, Iowa.

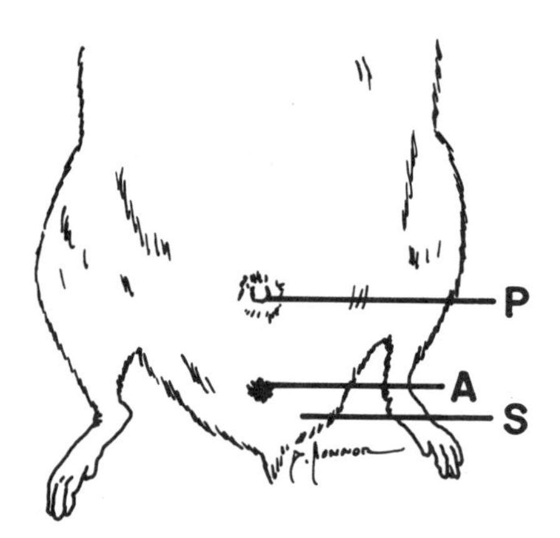

Fig. 17. External genitalia of male hamster. P = tip of penis; S = scrotal sac; A = anus. Adapted by permission from The Golden Hamster by R. A. Hoffman, P. F. Robinson, and H. Magalhaes, (c) 1968 by the Iowa State University Press, Ames, Iowa.

BREEDING

Sexing a Hamster

The perineal margin of the male hamster is rounded due to the scrotal sacs (Fig. 17), and the female posterior is pointed toward the tail (Fig. 18). Males have a greater anogenital distance and a pointed genital papilla containing a rounded penile opening.

Breeding Program

Hand Mating. The female hamster is placed into the male's cage for a few hours or a week, and the pair is observed for mating or fighting. The male is then removed. As males are frequently injured by females, aggressive males are more successful in the hand-mating scheme.

Harem Mating. One to 4 males are placed with 1 to 5 times as many females. Seven to 12 days after pairing the females are removed and individually housed. Fighting occurs with this mating system.

Monogamous Pair Mating. The male and female are paired at or before weaning and are left together. Fighting and destruction of the litter may occur with this system.

Young hamsters are weaned at 3 to 4 weeks.

Golden hamsters are bred for the first time at 6 to 8 weeks or at a weight of 80 to 100 gm.

Estrous Cycle

As estrus approaches in a sexually mature female, a slight, stringy, translucent mucus is extruded from the vagina. The morning following estrus a tenacious, opaque vaginal discharge appears. A receptive, nonbelligerent approach of the female to the male hamster is evidence of impending estrus.

Hamsters have a 4-day estrous cycle. An anovulatory postpartum estrus occurs. A fertile estrus follows weaning by several days.

Gestation Period in Hamsters

The hamster gestation period lasts from 15 to 18 days with a 16 day average.

Determination of Pregnancy

The discharge of estrus ceases at 5 and 9 days after mating and the observation of a distended abdomen or rapid weight gain are evidences of pregnancy.

Pseudopregnancy in Hamsters

Infertile matings or crowding of the females may produce an occasional pseudopregnancy of 7 to 13 days.

Nest Building

Hamsters, pregnant or not, build nests, but more nests are built when temperatures are cool. As neonatal hamsters are very immature and require a warm environment, the female builds and utilizes a maternal nest.

Size of Litter

A hamster litter usually contains from 5 to 10 young.

Abandonment and Cannibalism

Litter abandonment and cannibalism, common during the first week postpartum, may occur if the nest is disturbed, the colony is startled, the young bite the mother, the young are born on wire, agalactia or mastitis occurs, the diet is inadequate, or the litter is large. The young should not be handled until 10 days postpartum, especially if the female hamster is primiparous. At 14 days postmating the female should be supplied with 5 days' feed and bedding. This allows the mother to nest and remain undisturbed.

Hand Rearing

Fostering is rarely successful with hamsters, as both the adopted and natural litters may be cannibalized. Neonatal hamsters are immature and hand rearing is extremely difficult, if not impossible.

DISEASE PREVENTION

The general husbandry and disease prevention measures mentioned in Chapter 1 should be observed. Among major husbandry problems with hamsters are fighting, cannibalism of young, escape from cages, and susceptibility to climatic changes.

PUBLIC HEALTH CONCERNS

Lymphocytic choriomeningitis (LCM) has received attention because of a recent outbreak in man, but the severe forms of LCM are rare, even in humans at risk. Salmonellosis, tularemia, and *Hymenolepis* infections are hamster diseases of potential public health significance.

ANATOMIC AND PHYSIOLOGIC CHARACTERISTICS

Hibernation. Under certain conditions of lower temperatures (below 20° C) and decreased food supply and photoperiod, hamsters may hibernate for short periods.

Dental Caries. Caries have been experimentally produced in the golden hamster.

Long Renal Papilla. The renal papilla extends into the ureter, which allows the nephron, specifically the distal tubule, to be catheterized.

Reversible Cheek Pouches. These highly vascularized, thin epithelial tissues have been utilized in studies of microcirculation and, as the pouches are immunologically privileged sites, tumor transplantation.

Neoplasia. Hamsters are susceptible to a wide range of oncogenic viruses (SV5, polyoma, Rous sarcoma) and transplanted neoplastic tissue.

Concealed Young. The female hamster when excited is able to safely conceal her entire newborn litter in her cheek pouches. Later, when the perceived danger passes, the mother returns the young to the nest. Females may also conceal feed or bedding in their cheek pouches.

BIBLIOGRAPHY

Chen, K. K., Powell, C. E., and Maze, N.: The response of the hamster to drugs. J. Pharmacol. Exp. Ther., *85*:349-355, 1945.

Granados, H.: Nutritional studies on growth and reproduction of the golden hamster (*Mesocricetus auratus*). Acta Physiol. Scand., *24*:1-138, 1951.

Grindeland, R. E., Folk, G. E., and Fomand, R. L.: Some factors influencing the life span of golden hamsters. Proc. Iowa Acad. Sci., *64*:638-642, 1957.

Hoffman, R. A., Robinson, P. F., and Magalhaes, H.: The Golden Hamster. Ames, The Iowa State University Press, 1968.

Menzies, J. D. E.: Hamster husbandry and handling. J. Anim. Tech. Assoc., *12*:2-7, 1961.

Paup, D. C., Coniglio, L. P., and Clemens, L. G.: Hormonal determinants in the development of masculine and feminine behavior in the female hamster. Behav. Biol., *10*:353-363, 1974.

Schimidt, R. E.: Selected Topics in Laboratory Animal Medicine, Vol. XXIV: The Hamster. Brooks Air Force Base., Texas, USAF School of Aerospace Medicine, 1975.

Soderwall, A. L., and Britenbaker, A. L.: Reproductive capacities of different age hamsters (*Cricetus auratus*). J. Gerontol., *10*:469-470, 1955.

Waddell, D.: Hoarding behavior in the golden hamster. J. Comp. Physiol. Psychol., *14*:383-388, 1951.

The Gerbil

The Mongolian gerbil, *Meriones unguiculatus*, is a curious, nearly odorless, friendly rodent distinguished by its monogamous mating behavior, water conservation mechanisms, spontaneous epileptiform seizures, and paucity of common, spontaneous diseases.

TAXONOMY

Taxonomic Position of the Gerbil

Class:	Mammalia
Order:	Rodentia
Family:	Cricetidae
Genus:	Meriones
Species:	unguiculatus

Meriones unguiculatus, the Mongolian gerbil, is native to the arid regions of northern China and Mongolia. Gerbils have been classified with the Muridae and, more lately, with the Cricetidae. Eleven to 15 genera exist, depending on which taxonomist is referenced. Although other genera have rarely been used as research subjects in the United States, in other countries *M. libycus*, *M. shawii*, and *M. tristami* have been used for certain studies. Gerbillinae are fairly widespread throughout Africa, Eastern Europe, and the Middle East, and Asia.

Varieties of Gerbils Available

The agouti or wild-color Mongolian gerbil is the only variety of *Meriones unguiculatus* commonly available.

SOURCES OF INFORMATION

The Gerbil Digest, published quarterly by Tumblebrook Farm, Inc., West Brook-

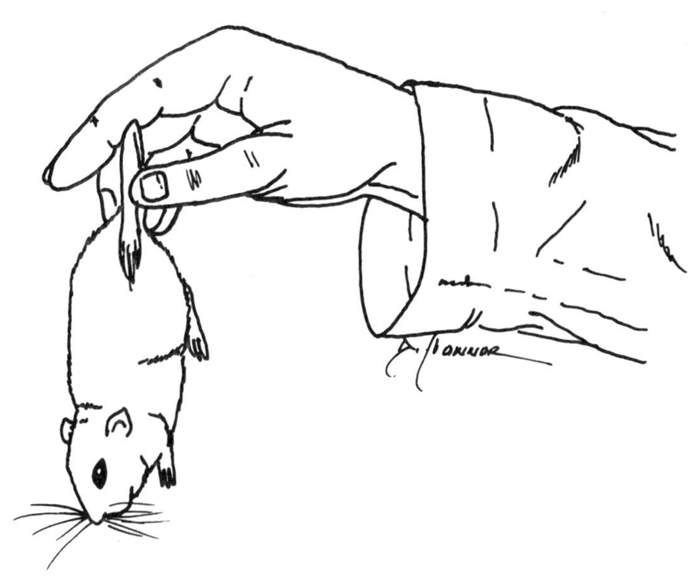

Fig. 19. Picking up a gerbil with a grip at the tail base. If the tail of a rodent is grasped distal to the base, the tail skin may be pulled off.

field, MA 01585, is one of several publications on gerbils available.

Enjoy Your Gerbil and *Enjoy Your Gerbils, Rats, and Mice* are available in pet stores or from The Pet Library, Ltd., 600 South 4th Street, Harrison, NJ 07029.

How to Raise and Train Gerbils (1967) by D. G. Robinson, Jr., is available from T. F. H. Publications, Inc., P.O. Box 33, Jersey City, NJ 07303.

The Aeromedical Review (7-74) Volume XXIII, *The Mongolian Gerbil,* is distributed by the National Technical Information Service, 5285 Port Royal Road, Springfield, VA 22161.

GERBILS AS PETS

Gerbils are clean, friendly and curious, quiet, leave little odor, rarely bite, and are easily handled. However, because of the possibility of gerbils becoming established in the wild, they are banned as pets in California.

Life Span of Gerbil

The reproductive life of a female gerbil is approximately 18 months, but the life span may exceed 3 years.

Commercial Market for Gerbils

Pet stores, private individuals, and research institutions purchase gerbils, and several commercial suppliers exist. The potential market should be explored before any investment in animals and facilities is made.

Picking Up and Carrying a Gerbil

The gerbil's tail is grasped at the base, and the animal lifted and cradled in the hand (Fig. 19). The gerbil may be more securely restrained with a grip over the back (Fig. 20). Gerbils resist being placed on their backs and, while struggling, may accidently drop to the ground.

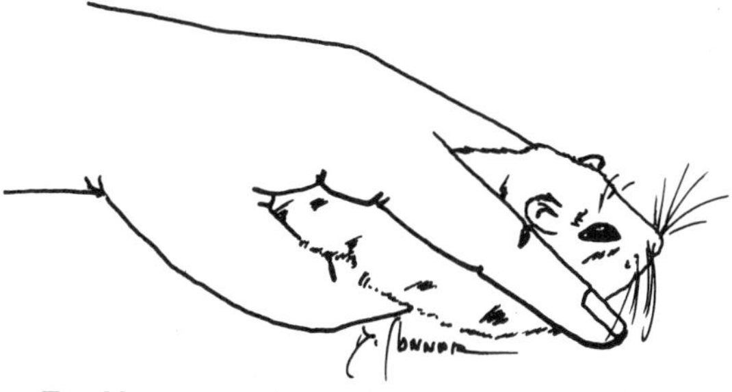

Fig. 20. An over-the-back grip for restraining a gerbil.

Fighting

Gerbils paired after reaching puberty (10 wk) or reunited after prolonged separation may fight. Fighting is reduced if the gerbils reach maturity in the same cage or if the animals are allowed to recover simultaneously from anesthesia in a neutral cage.

HOUSING

Plastic or metal rat or mouse cages, with a solid bottom and bedding material, are the usual housing arrangement. Bedding should be of sufficient depth (2 cm) to facilitate nest building and should not be of an abrasive composition. Provision of an opaque cage or a hiding place within the cage improves reproductive and maternal performance. Gerbils are active gnawers and burrowers. Cages should be designed to prevent escape.

Gerbils over 12 weeks of age require a floor area of 232 cm^2 (36 in^2) per animal. A breeding pair requires a cage area of 1290 cm^2 (200 in^2). Cage sides should be at least 15 cm (6 in) high.

Gerbils are housed at temperatures between 18° and 29° C (65° to 85° F) with the median ambient temperature at 24° ± 1° C (75° ± 2° F). The environmental humidity should be above 30% saturation, although at approximately 50% and above the gerbil's hair coat becomes roughened.

Outdoor Housing

Small laboratory or pet rodents are not housed outdoors. Climatic extremes impair health and reproductive performance. Escape from the cage is always a possibility.

Cleaning Gerbil Cages

Gerbils produce only a few drops of urine daily, and their bedding remains odorless for several weeks. In gerbil colonies the bedding is usually changed when dirty or every 2 weeks. Cages should be washed with hot water and detergent, disinfected, and rinsed.

FEEDING AND WATERING

Gerbils should be fed *ad libitum* a complete, freshly milled, pelleted rodent chow. Gerbils prefer sunflower seeds to the pelleted chows, but these seeds, with a low calcium and high fat content, are not a complete diet. Gerbils 2 to 5 weeks old may have difficulty opening the sunflower seeds or gnawing the hard pellets. Although gerbils in the wild state require little water, caged gerbils should be supplied continuously with clean water.

Adult gerbils will consume approximately 7 gm feed and 4 ml water daily. Rodent chows contain approximately 24% crude protein and 5% crude fiber.

BREEDING

Sexing a Gerbil

Young males have a dark scrotum, and the anogenital distance is greater in males (10 mm) than in females (5 mm). Both sexes have a genital papilla (Figs. 21 and 22).

Breeding Program for Gerbils

Polygamous harem groups are successful if established before the gerbils are sexually mature. Compatible monogamous pairs are formed at 8 weeks of age or earlier and should not be separated unless fighting or rejection occurs. Some breeders remove the male for 2 weeks postpartum to reduce the disturbance to the female and young. The young are weaned at 3 to 3.5 weeks.

Gerbils are mated at 10 to 12 weeks, when they weigh approximately 70 to 80 gm. Gerbils should be paired before they reach sexual maturity.

Estrous Cycle

As gerbils are permanently paired, the detection of estrus by the handler is not a

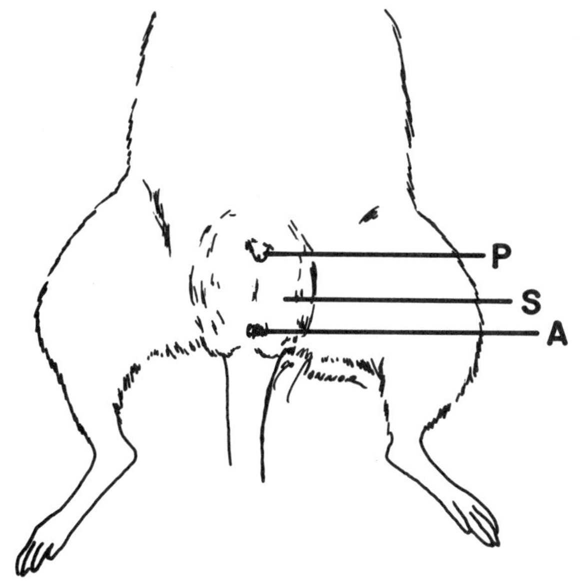

Fig. 21. External genitalia of the male gerbil. P = tip of penis; S = scrotal sac; A = anus.

factor in breeding. Gerbils in estrus act restless and may have a congested vulva.

The gerbil estrous cycle lasts 4 to 6 days. A fertile, postpartum estrus occurs in a majority of gerbils.

Gestation Period in Gerbils

The gestation period in gerbils lasts 24 to 26 days.

Determination of Pregnancy

During pregnancy a mature gerbil will gain between 10 and 30 gm. There are inconsistent reports of blood in the vaginal smear from 13 days postmating until parturition.

Pseudopregnancy in Gerbils

Pseudopregnancy occurs in gerbils but is uncommon. In gerbils pseudopregnancy lasts 13 to 23 days.

Nest Building

Gerbils, pregnant or not, build nests in the bedding. This activity is accentuated in cooler temperatures. Nests are constructed of bedding materials, which gerbils are adept at shredding. Maternal nests are often covered. A wooden box 12 cm on a side placed in the cage facilitates nest building.

Size of Litter

Usually 4 to 6 gerbils are born at one time; smaller litters may be destroyed by the mother. If the litter is destroyed and lactation ceases, the female resumes cycling. There may be a selective advantage in the wild for destroying smaller litters and rebreeding.

Abandonment and Cannibalism

Young are rarely abandoned or cannibalized. Factors contributing to abandonment include small litters (3 or fewer young), excessive handling of the young, lack of nesting material, and cages without provision for concealment.

Hand Rearing

Fostering is possible if the orphaned and host litters are within a few days of each other. Hand feeding of neonatal rodents is difficult.

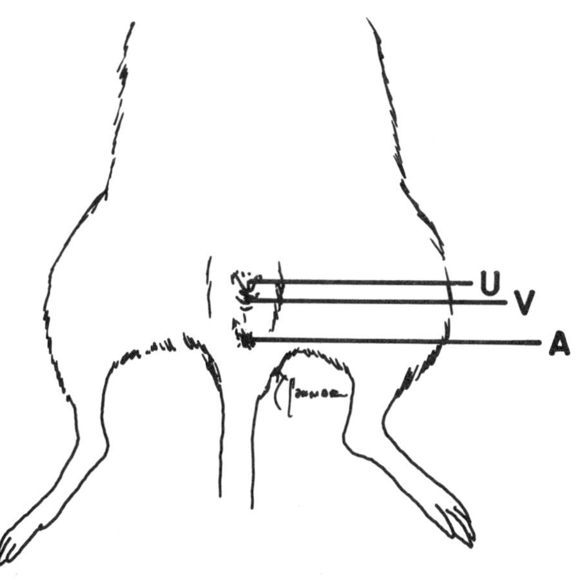

Fig. 22. External genitalia of the female gerbil. U = urethral orifice; V = vaginal orifice; A = anus.

DISEASE PREVENTION

Observe the general husbandry admonitions listed in Chapter 1. The burrowing activities of gerbils often lead to ulceration of the nose. The fondness of gerbils for sunflower seeds can produce the false impression of good dietary habits; however, sunflower seeds are not a complete diet for rodents.

PUBLIC HEALTH CONCERNS

Gerbils in captivity have few spontaneous diseases and even fewer of public health significance. *Salmonella* and *Hymenolepis* infections are potential health problems for man. Both conditions are rare in gerbil colonies.

ANATOMIC AND PHYSIOLOGIC CHARACTERISTICS

Maze Running. Although gerbils learn avoidance responses quickly, their curiosity and seeking behaviors make gerbils poor maze runners.

Epileptiform Seizures. Spontaneous, convulsive seizures occur in approximately 20% of gerbils. There is a gradation from mild to severe in the seizure pattern.

Radiation Tolerance. The LD_{50} X or gamma radiation dose in gerbils is approximately 1200R compared to approximately 600R in other laboratory species.

Incomplete Circle of Willis. Some gerbils develop cerebral ischemia following unilateral carotid ligation.

Adrenal Glands. The adrenal glands of the gerbil are large in weight relative to the whole body weight.

High Serum Cholesterol. Even on diets with standard range fat levels (2% to 4%), gerbils have, compared to other laboratory species, high blood serum cholesterol levels. The gerbil is a naturally lipemic animal.

Renal System. The nephron is remarkably adapted to conserve water and electrolytes.

BIBLIOGRAPHY

Arrington, L. R., and Ammerman, C. B.: Water requirements of gerbils. Lab. Anim. Care, *19*:503-505, 1969.

Arrington, L. R., Beaty, T. C., and Kelley, K. C.: Growth, longevity, and reproductive life of the Mongolian gerbil. Lab. Anim. Sci., *23*:262-265, 1973.

Chang, M. C., Hunt, D. M., and Turbyfill, C.: High resistance of Mongolian gerbils to irradiation. Nature, *203*:536-537, 1964.

David, T. D.: Selected Topics in Laboratory Animal Medicine, Vol. XXIII: The Mongolian Gerbil. Brooks Air Force Base, USAF School of Aerospace Medicine, 1974.

Glickman, S. E., Fried, L., and Morrison, B. A.: Shredding of nesting material in the Mongolian gerbil. Percep. Mot. Skills, *24*:473-474, 1967.

Kuehn, R. E., and Zucker, I.: Reproductive behavior of the Mongolian gerbil (*Meriones ungiuculatus*). J. Comp. Physiol. Psychol., *66*:747-752, 1968.

Levine, S., and Sohn, D.: Cerebral ischemia in infant and adult gerbils. Arch. Pathol., *87*:315-317, 1969.

McManus, J. J., and Mele, J. A.: Temperature regulation in the Mongolian gerbil, *Meriones unguiculatus*. Bull. N. J. Acad. Sci., *14*:21-22.

Marston, J. H., and Chang, M. C.: The breeding, management and reproductive physiology of the Mongolian gerbil, *Meriones unguiculatus*. Lab. Anim. Care, *15*:34-48, 1965.

Mitchell, O. G.: The supposed role of the gerbil ventral marking gland in reproduction. J. Mammal., *48*:142, 1967.

Nakai, K., et al.: Reproduction and postnatal development of the colony-bred *Meriones unguiculatus*. Bull. Exp. Animals (Japan), *9*:157-159, 1960.

Pav, D. I., and Magalini, S. I.: Studies on the *Meriones unguiculatus* (Mongolian gerbil). 1. Observations on history, anatomy, and laboratory habits. Metabolism, *2*:139-149, 1966.

Rich, S. T.: The Mongolian gerbil (*Meriones unguiculatus*) in research. Lab. Anim. Care, *18*:235-243, 1968.

Robinson, P. F.: Metabolism of the gerbil, *Meriones unguiculatus*. Science, *130*:502-503, 1959.

Roscoe, H. G., and Fahrenback, M. J.: Cholesterol metabolism in the gerbil. Proc. Soc. Exp. Biol. Med., *110*:51-55, 1962.

Schwentker, V.: Care and Maintenance of the Mongolian Gerbil. West Brookfield, Mass., Tumblebrook Farm, Inc., 1967.

Seiler, M. W., et al.: Hyperlipemia in the gerbil: Effect of diet on hepatic lipogenesis. Am. J. Physiol., *221*:554-558, 1971.

Walters, G. C., Pearl, J., and Rogers, J.: The gerbil as a subject in behavioral research. Psychol. Rep., *12*:315-318, 1963.
Williams, W. M.: The Anatomy of the Mongolian Gerbil. West Brookfield, Mass., Tumblebrook Farm, Inc., 1974.
Winkelmann, J. R., and Getz, L. L.: Water balance in the Mongolian gerbil. J. Mammal., *43*:150-154, 1962.

The Mouse

The domestic or house mouse, *Mus musculus,* is occasionally encountered as a pet, but over 27 million mice are used annually in medical research in the United States. Mice are prolific breeders, easily maintained in large populations, possess great genetic diversity, and are well characterized anatomically and physiologically.

TAXONOMY

Taxonomic Position of the Mouse

Class:	Mammalia
Order:	Rodentia
Family:	Muridae
Genus:	Mus
Species:	musculus

Mus musculus, the house mouse, evolved in the Old World and has lived for millennia in areas of human habitation. The breeding of mice by fanciers led to the genetic diversity of the mouse population and the research interest of the nineteenth century scientists. The Swiss albino mouse is the source of many noninbred, white laboratory mice, although there are several strains of albino or white inbred mice. Several genera of wild mice are encountered in research colonies, including the field mouse *Microtus,* the grasshopper mouse *Onychomys,* and the white-footed mouse *Peromyscus.*

Types of Mice Available

Mice may be included in two major categories based on their ecologic and genetic characteristics.

The ecologic category includes (1) germ-free (axenic) mice, which are free from detectable microorganisms; (2) defined flora (gnotobiotic) mice, which possess a specified flora and fauna; (3) specific pathogen-free mice, which are free of specified pathogenic microorganisms; and (4) conventional mice, which are mice not included in the other categories. Other laboratory animals may also be classified ecologically.

The genetic category includes (1) random-bred mice, where, in a large colony, matings occur randomly among males and females from unrelated litters and different cages; (2) inbred mice, the genetically homogeneous products of at least 20 consecutive brother-sister matings; and (3) the F_1 hybrid offspring of two lines of inbred parents. Several inbreeding schemes exist. Each brother-sister mating reduces existing heterozygosity by approximately 19%. Selective breedings and inbreedings have produced the great diversity of mice encountered as pets and research animals.

Inbred mice come in a wide variety of attractive, strain-specific coat colors. Consequently, these strains find their way into laboratories, schools, and homes where they are bred and raised as pets rather than as the model of human disease, that characteristic for which they were originally selected. Regardless of their use, these mice are destined to develop those complex disease conditions characteristic of the strain. This variation in strain susceptibilities presents the small animal practitioner with multiple differential diagnostic challenges.

SOURCES OF INFORMATION

The United States Department of Agriculture publishes Leaflet No. 483 (1961) *Raising Mice and Rats for Laboratory Use.* This leaflet is available for 5 cents from the Superintendent of Documents, United States Government Printing Office, Washington, DC 20402.

A paperback book, *The Laboratory Mouse,* by M. L. Simmons and J. O. Brick, has been published (1970) by Prentice-Hall, Inc., 301 Sylvan Avenue, Englewood Cliffs, NJ 07632.

The USAF School of Aerospace Medicine, Brooks Air Force Base, publishes a booklet entitled *The Mouse,* Vol. XIX, in the Selected Topics in Laboratory Animal Medicine series. The booklet is available from the National Technical Information Service, 5285 Port Royal Road, Springfield, VA 22161.

The standard reference work on the mouse, available in medical libraries, is *The Biology of the Mouse,* 2nd Edition (1966), E. L. Green, Editor. This text was published by McGraw-Hill, New York and London, in 1966.

The Anatomy of the Laboratory Mouse, by M. J. Cook, was published (1965) by Academic Press, Inc., 111 Fifth Avenue, New York, NY 10003.

MICE AS PETS

Mice, if handled gently, make good, albeit small, pets. Mice are timid, social, nocturnal, and escape-prone rodents that require little cage space and small quantities of feed and water. Mice housed individually tend to become aggressive. Human allergies to mouse dander are common. Mice may have an undesirable odor, particularly if male mice are maintained or the cages are excessively moist and dirty.

Fig. 23. Restraint of a mouse by grasping the tail base.

Life Span of a Mouse

Mice may live from 2 to 3 years. There are major longevity differences among strains, owing mainly to differential disease susceptibilities. For example, AKR and C3H mice live shorter lives than A and C57BL strains. Mice breed for approximately 12 to 18 months (6 to 10 litters).

Commercial Market for Mice

As with all commercial laboratory animal endeavors, the market should be explored before production begins. Mice, because of their small size, fecundity, ease of handling, low individual cost, and genetic diversity, comprise approximately 70% of the 36 million animals used annually in biomedical research in the United States.

Picking Up and Carrying a Mouse

Mice may be picked up by grasping the tail with the fingers (Fig. 23) or the body with rubber-tipped forceps. If inspection or manipulations are intended, the mouse is lifted by the tail, placed on a rough, "toe-gripping" surface, grasped on the scruff of the neck by thumb and forefinger, inverted, and the tail is held between the little finger and palm (Fig. 24). Grasping and pulling the tip of the tail may result in stripping the skin from the tip of the tail.

Fighting

Two or more adult male mice housed together are usually incompatible; therefore, male mice should be housed separately to prevent fighting with resulting

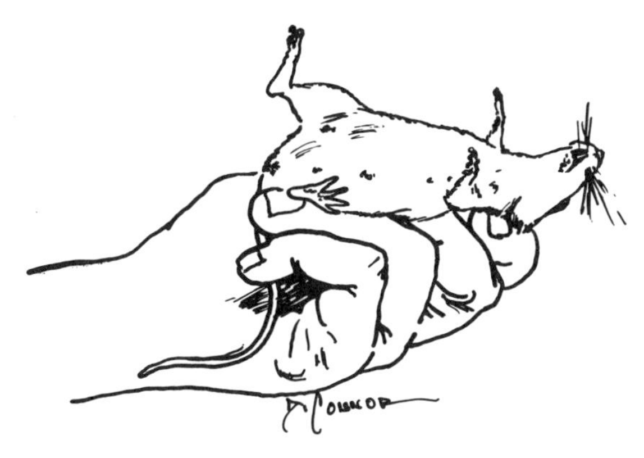

Fig. 24. One-hand restraint of a mouse for injection or bleeding procedures.

abscesses, dermatitis, septicemia, and death. Newly assembled male groups, new males entering established territories, or mice previously housed alone are more susceptible to fighting, although female mice seldom fight.

HOUSING

Mice used in research are housed either in metal or plastic shoe-box type cages with wire or slotted tops or in suspended wire cages. "Mouse mesh," whether used for research or pet caging, should have openings of 0.65 cm sq or 3 wires per inch. A homemade shoe-box cage with a hardware cloth top is satisfactory for pet mice, as are the several wire cages available in pet stores. Large mice will have difficulty eating through 0.65 cm mesh, and weanling mice may escape through larger meshes. Care should be taken to prevent escape, loss of young, and fractured limbs.

Mouse cages should provide at least 97 cm^2 (15 in^2) floor space per adult (30 gm) mouse. A female with litter should have 390 cm^2 (63 in^2) floor space. Mouse cages should be at least 13 cm (5 in) high.

Bedding, which may be paper, soft pine, aspen, or cedar wood shavings, peat moss, beet or alfalfa pulp, or sawdust, should be nonallergic, dust-free, absorbent, nontoxic, and free of pathogens. Soft wood shavings make excellent rodent nesting material.

Temperatures in mouse rooms should be maintained between 18° and 29° C (65° to 85° F), the median ambient temperature should be maintained at 24° ± 1° C (75° ± 2° F), and the humidity should be between 30% and 70% saturation. The use of cage filter covers raises ambient temperature and humidity and makes their control more difficult. Mice are adept at gnawing through and escaping from cages. Mice housed alone tend to become more aggressive.

Outdoor Housing

Because mice have a tendency to escape and are vulnerable to predators and environmental stresses, they are not housed outdoors.

Cleaning Mouse Caging

Mouse caging, like all animal caging and bedding, is cleaned or changed as often as necessary to prevent accumulation of odor and waste and to keep the animals dry and clean. Caging is usually cleaned at least twice a week by the washing or disinfection methods described in Chapter 1.

FEEDING AND WATERING

Mice are fed a clean, wholesome, and nutritious pelleted rodent chow *ad libitum* and watered with an automatic watering system or water bottles with sipper tubes. Several varieties of special purpose rodent chows are available. Many of the pet rodent feeds found in pet stores are poor diets, primarily because of excessive fiber and low protein content.

Rodent chows used for rats, mice, hamsters, and gerbils contain approximately 24% crude protein and 4% crude fiber. An adult mouse will consume approximately 12 gm feed per 100 gm body weight per day. Water and feed consumption varies with the ambient temperature, humidity, dryness of the feed, breeding status, feed quality, and state of health.

BREEDING

Sexing a Mouse

Neonatal males may be distinguished from females by a greater anogenital distance in males (1.5 to 2X), the pale testes visible through the abdominal wall, the larger genital papilla, and in females at 9 days the conspicuous rows of nipples (Figs. 25 and 26).

Breeding Program for Mice

Considerations in mouse breeding programs include available space, strain fecundity, inbreeding schemes, epidemiologic concerns, and production requirements. Mice are continuously polyestrus with minor seasonal variations. Mice have a fertile, postpartum estrus occurring 14 to 28 hours after parturition, and some breeding schemes utilize this estrus for increased production. Simultaneous lactation and pregnancy, however, may delay implantation 3 to 5 days.

Young mice are weaned at 18 to 21 days (10 to 12 gm). If the postpartum estrus is not utilized, the female resumes cycling 2 to 5 days postweaning.

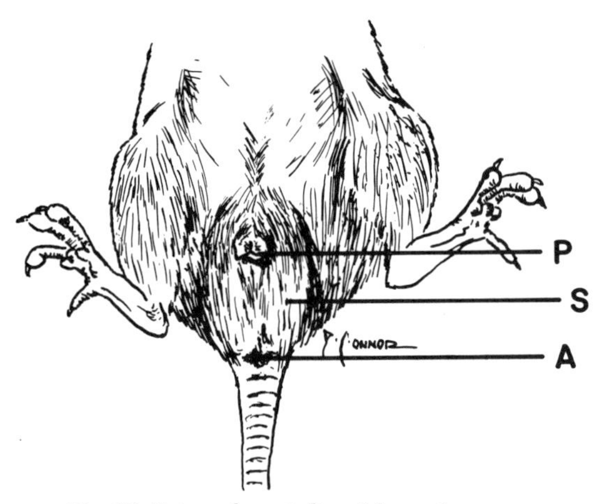

Fig. 25. External genitalia of the male mouse. P = tip of penis; S = scrotal sac; A = anus.

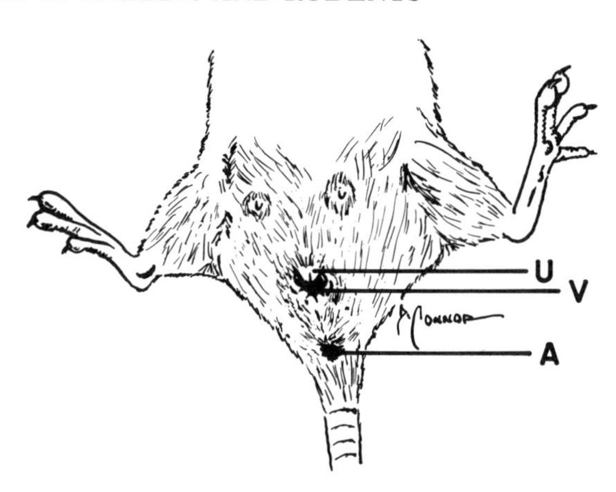

Fig. 26. External genitalia of the female mouse. U = urethral orifice; V = vaginal orifice; A = anus.

Representative breeding systems for mice include the colony, the monogamous, and polygamous mating schemes. In colony mating 1 male and 2 to 6 females are housed together continuously, and the young are removed at weaning. This system is the most efficient of all for space and labor utilization, but record keeping is difficult.

The monogamous system involves the constant pairing of one male and one female. The young are removed prior to the next parturition. This system, which utilizes the postpartum estrus, produces the maximum number of litters in the shortest time and provides for ease of record keeping and evaluation of individual female production. Disadvantages of monogamous breeding include the large male population and the increased need for labor, space, and equipment to service the increased populations.

The polygamous or harem scheme combines 1 male and 2 to 6 females. Females are removed to separate cages prior to parturition; the postpartum estrus is not utilized. In this system each female provides more milk, larger young, and more young weaned per litter. Disadvantages include lowered total number of litters, increased labor time per cage, and possibly male exhaustion and pseudopregnancies. The cross-fostering modification combines 1 male and 4 females. Each female remains with the group until 24 hours postpartum for breeding. She is then removed (with litter) to a nursing cage with 3 other recently whelped mothers. At weaning, nonpregnant females are returned to the male.

Although mice may have an estrus at 35 to 50 days, they are usually first bred when 50 days of age (20 to 30 gm). Mice bred too early or too late (after 10 wk) have reduced fertility. The onset of sexual maturity varies with strain and growth rate, litter size, and level of nutrition.

Receptivity for Mating

The short estrous cycle and the utilization of the postpartum estrus reduce the necessity for cycle monitoring. During proestrus the vaginal smear contains epithelial cells with some cornified cells and leukocytes; in estrus cornified cells become prominent. In late metestrus and diestrus cornification decreases and lymphocytes increase.

Estrous Cycle in Mice

The estrous cycle lasts 4 to 5 days, with an average evening estrous period of 12 hours. Except for the postpartum estrus, estrus does not occur during lactation. Group-caged female mice may enter a period of continuous anestrus, which is terminated by the odor or presence of a male. The group of females will then come into estrus in approximately 72 hours. This synchronization of estrus is known as the Whitten effect. The infrequent pseudopregnancies may cause an increased interval between estrous periods.

Gestation Period in Mice

In the nonsuckling dam the gestation period ranges from 19 to 21 days. Simultaneous lactation and pregnancy prolong the period 3 to 10 days, depending on the size of the suckling litter. The gestation period in hybrid mice tends to be shorter than the period in inbred mice. If a mouse bred within the previous 24 hours is exposed to a strange male, the existing pregnancy will be aborted. This phenomenon is known as the Bruce effect.

Determination of Pregnancy

Detection of sperm by vaginal examination or the presence of a coagulation-copulatory plug in the vagina are evidences of mating within the past 24 hours. As gestation proceeds, the mammary structures develop, and the fetuses can be palpated. Daily weighing will reveal an increased rate of weight gain about 13 days gestation. Mammary development is pronounced at 14 days.

Pseudopregnancy in Mice

Pseudopregnancy in mice is rarely noticed but may follow male exhaustion and sterile matings. The pseudopregnancy may last 1 to 3 weeks.

Nest Building

Mice routinely prepare small sleeping nests in the bedding. The brood nest, which may contain one or more families, is a large, hollow nest prepared in late gestation. Litter fostering occurs. The female spends much time with the young in the nest. Mice will retrieve young.

Size of Litter

Litter size varies considerably with strain and age. The first litter is generally smaller, with optimal production (10 to 12 young) occurring between the second and eighth litters. Inbred strains usually produce smaller litters.

Abandonment and Cannibalism

Cannibalism is uncommon in mice, but female mice whelping or recently whelped should remain undisturbed for at least 2 days postpartum.

Hand Rearing

Hand rearing and foster nursing underlie germ-free and gnotobiotic mouse production. These elaborate techniques are better left to the specialized breeding colony.

DISEASE PREVENTION

In addition to the routine husbandry precautions listed in Chapter 1, a systematic diagnostic evaluation of colony animals should routinely be conducted to screen for subclinical infections. Filter cage covers prevent or reduce airborne transmission of microorganisms within a densely populated mouse room, although the use of such covers increases the ammonia level, humidity, and temperature within the cage. Chlorination (10 to 12 ppm) or acidification (pH 2.5) of the drinking water will reduce the danger of *Pseudomonas* contamination. A solution with a pH of 2.5 is prepared by mixing 10 liters water with approximately 2.6 ml concentrated hydrochloric acid. Because of variations in the pH of local water supplies, the pH of the final mixture should be monitored.

PUBLIC HEALTH CONCERNS

Salmonellosis and lymphocytic choriomeningitis are zoonotic diseases of rare occurrence in mouse colonies. Rabies vaccinations are not usually indicated for persons bitten by caged laboratory rodents. The decision to use a killed rabies vaccine to protect a pet or research rodent or rabbit depends on considerations of the exposure to the wild animal population.

ANATOMIC AND PHYSIOLOGIC CHARACTERISTICS

The standard white mouse is an undistinguished, well-known small rodent noted more for its small size, rapid heart rate (600/min), high oxygen consumption (1.7 ml/gm/hr), and fecundity rather than for any unusual characteristic. However, over 60 years of inbreeding and strain development have produced a considerable reservoir of mutants, which are utilized by the millions in the investigations of a vast array of abnormalities. The following characteristics provide a small sample of the available mutants: anemic, athymic, bent tail, diabetic, frizzy, fuzzy, grizzled, jolting, naked, nervous, radiation sensitive, and tottering. A wide variety of spontaneous neoplasms occurs in several inbred strains.

Listings of available inbred strains can be obtained from the National Institutes of Health's *Catalogue of NIH Rodents*, NIH, Bethesda, MD 20014 (DHEW Publication NH-74-606) or from the Roscoe B. Jackson Memorial Laboratory, Bar Harbor, ME 04609.

BIBLIOGRAPHY

Brick, J. O., Newell, R. F., and Doherty, D. G.: A barrier system for a breeding and experimental rodent colony: description and operation. Lab. Anim. Care, *19*:92-97, 1969.

Bruce, H. M., Land, R. B., and Falconer, D. S.: Inhibition of pregnancy block in mice by handling. J. Reprod. Fertil., *15*:289-294, 1968.

Eisen, E. J.: Comparison of two cage rearing regimes on reproductive performance and body weight of the laboratory mouse. Lab. Anim. Care, *16*:447-453, 1966.

Green, E. L.: Biology of the Laboratory Mouse, 2nd Ed. New York and London, McGraw-Hill, 1966.

Hickey, T. E., and Tompkins, E. C.: Housing mice in a caging system with automatic flushing. Lab. Anim. Sci., *25*:289-291, 1975.

Iturrian, W. B., and Fink, G. B.: Comparison of bedding material; habitat preference of pregnant mice and reproductive performance. Lab. Anim. Care, *18*:160-164, 1968.

Knapka, J. J., Smith, K. P., and Judge, F. J.: Effect of open and closed formula rations on the performance of three strains of laboratory mice. Lab. Anim. Sci., *24*:480-487, 1974.

Les, E. P.: Cage population density and efficiency of feed utilization in inbred mice. Lab. Anim. Care, *18*:305-313, 1968.

North, W., et al.: A comparison of individual litter and colony rearing of mice. Lab. Anim. Care, *17*:408-412, 1967.

Pick, J. R., and Little, M. J.: Effect of type of bedding material on thresholds of pentylenetet-

razol convulsions in mice. Lab. Anim. Care, *15*:29-33, 1965.
Sigdestad, C. P., et al.: Evaluating feeding cycles of small rodents. Lab. Anim. Sci., *24*:919-923, 1974.
Simmons, M. L., et al.: Effect of a filter cover on temperature and humidity in a mouse cage. Lab. Anim., *2*:113-120, 1968.
Simmons, M. L., and Brick, J. O.: The Laboratory Mouse: Selection and Management. Englewood Cliffs, N.J., Prentice-Hall, Inc., 1970.
Smith, M. J., and Ryle, M.: An approach to the more economical production of uniform mice. J. Inst. Anim. Tech., *19*:74-77, 1968.
Spalding, J. F., Akchuleta, R. F., and Holland, L. M.: Influence of the visible color spectrum on activity in mice. Lab. Anim. Care, *19*:50-54, 1969.
Stoddart, R. C.: Breeding and growth of laboratory mice in darkness. Lab. Anim., *4*:13-16, 1970.
Vessell, E. S.: Induction of drug metabolizing enzymes in liver microsomes of mice and rats by soft wood bedding. Science, *157*:1057-1058, 1967.
Weltman, A. S., et al.: Effects of isolation stress on female albino mice. Lab. Anim. Care, *18*:426-435, 1968.
Whitten, W. K.: Modification of the estrus cycle of the mouse by external stimuli associated with the male. J. Endocrinol., *13*:399-404, 1956.
Zakem, H. B., and Alliston, C. W.: The effects of noise level and elevated ambient temperatures upon selected reproductive traits in female Swiss-Webster mice. Lab. Anim. Sci., *24*:469-475, 1974.

The Rat

The domesticated variety of the wild brown or Norwegian rat, *Rattus norvegicus,* usually represented by the albino animal in research colonies, is an Old World import which occupies an important place in biomedical research. Rats are well-defined, easily maintained, inexpensive, relatively healthy, and suited for a wide range of research procedures.

TAXONOMY

Taxonomic Position of the Rat

Class:	Mammalia
Order:	Rodentia
Family:	Muridae
Genus:	Rattus
Species:	norvegicus

Rattus norvegicus was raised by fanciers in the seventeenth century and by the early 1900's had become a popular animal in medical and behavioral research. The black or roof rat, *Rattus rattus,* is less often encountered in research colonies. Native North American "rats," members of the family Cricetidae, include the cotton rat *Sigmondou* and the kangaroo rat *Dipodormys.*

Types of Rats Available

Rats, like mice, are available in various ecologic (germ-free, gnotobiotic, specific pathogen free, and conventional) and genetic varieties; however, most laboratory rat colonies contain conventional, random-bred animals with the albino or piebald (hooded) coat colors. Commonly available strains or varieties of rats include the Sprague-Dawley, the Wistar, and the Long-Evans rats.

The Sprague-Dawley rat, an albino, has a head narrower and a tail longer than the other varieties. This rat is more prolific and is claimed by some breeders to be less resistant to respiratory disease than the Wistar rat.

The Wistar rat, an albino, has a wide head, tail length shorter than body length, long ears, and may be more resistant to some infectious diseases than other rats, although such claims are often based on small or limited populations.

The Long-Evans and other "hooded" varieties are smaller than the albino strains and have darker hair over portions of the head and dorsal trunk.

SOURCES OF INFORMATION

The United States Department of Agriculture publishes leaflet 483 (1961) *Raising Mice and Rats for Laboratory Use.* This leaflet is available for 5 cents from the Superintendent of Documents, United States Government Printing Office, Washington, DC 20402.

The Rat in Laboratory Investigation, edited by E. J. Farris and J. Q. Griffith (1949), was reprinted in 1969 by the

Hafner Publishing Company, Inc., 31 East 10th Street, New York, NY 10003.

The Biology of the Laboratory Rat is an upcoming title in the series of reference texts edited by the American College of Laboratory Animal Medicine (ACLAM) and published by Academic Press, Inc., 111 Fifth Avenue, New York, NY 10003.

The Rat: A Study in Behavior by S. A. Barnett was revised (1975) by the University of Chicago Press, 5801 Ellis Ave., Chicago, IL 60637.

Anatomy of the Rat by E. C. Breene is published by Hafner Publishing Company, Inc., 31 East 10th Street, New York, NY 10003.

RATS AS PETS

Despite their association with disease, garbage, and sorcery, rats, if gently handled, make quiet, intelligent, easily trained, and gentle pets. Rats may escape from a cage but will return for food and water.

Fig. 27. Forequarters grip for restraining a rat.

Life Span of a Rat

Rats may live in excess of 3 years and have a productive breeding life until 14 months of age, during which time they may bear six to ten litters with 6 to 12 offspring per litter.

Commercial Market for Rats

Approximately 15 million rats are used annually in research, but the potential breeder should be aware that research investigators require rats with certain characteristics. Because many researchers require specific pathogen-free rats, large numbers of a particular strain and specified sex and age, and delivery, with little advance notice, on a specific date, it is difficult for individuals to profitably enter research animal production.

Picking Up and Carrying a Rat

Rats gently handled become tame and will rarely bite unless startled or hurt. Because the tail skin may tear, the tail should be grasped only at the base and for short periods. Rats in wire-bottomed cages, in particular, should not be grasped by the tail because the animal can hold on to caging.

Rats are picked up by placing the hand firmly over the back and rib cage and restraining the head with thumb and forefinger immediately behind the mandibles (Fig. 27). Rats held upside down are more concerned with righting themselves than with biting.

Fighting

Fighting is uncommon in rats, and, unlike mice, males can be housed in groups. Females also may be grouped, but postparturient females may fight among themselves. Isolation of rats does not produce the same degree of aggressive behavior toward other rats as seen with isolated mice toward other mice.

HOUSING

Rats used in research are housed either in metal or plastic shoe-box type cages with wire or slotted tops or in suspended, wire cages. "Rat mesh," whether used for research or pet caging, should have openings of 0.85 cm sq or 2 wires per inch. A homemade, shoe-box shaped cage with a hardware cloth top is satisfactory for pet rats, as are the several rodent cages available in pet stores. Care should be taken to prevent escape, loss of neonates, and fractured limbs.

Rat cages should provide at least 250 cm^2 (40 in^2) space per adult (300 gm) rat. A female with litter requires 1000 cm^2 (155 in^2) floor space. Rat cages should be at least 18 cm (7 in) high.

Bedding, which may be paper, soft wood shavings, ground corn cob, peat moss, beet or alfalfa pulp, or sawdust, should be nonallergic, dust-free, absorptive, nontoxic, and clean. Soft shavings are more easily formed into maternal nests.

Temperatures in rat rooms should be maintained between 18° and 29° C (65° to 85° F), with the median ambient temperature at 24° ± 1° C (75° ± 2° F), and the humidity should be between 30% and 70% saturation. The use of cage filter covers makes ambient temperature, ammonia, and humidity levels more difficult to control.

Outdoor Housing

The vulnerability of rats to predators, susceptibility to respiratory disease, potential for exposure to wild rodents, and escape tendencies make outdoor housing ill-advised.

Cleaning Rat Caging

Litter or bedding should be changed as often as necessary to keep odor minimal and the rats dry and clean; one to three bedding changes per week are usually sufficient. Cages, feeders, and water bottles are washed once or twice weekly. General husbandry concerns, described in Chapter 1, apply to facilities housing rats.

FEEDING AND WATERING

Rats are fed a clean, wholesome, and nutritious pelleted rodent chow *ad libitum* or in amounts that can be emptied from the feeder twice weekly. Rats should be watered from an automatic watering system or with water bottles and sipper tubes. Several varieties of special purpose rodent chows are available. Rats are cautious eaters (neophobic) and will avoid strange foods. Some of the feed found in pet stores is inadequate; food should be obtained from a supplier of research diets. Rodent chows used for rats, mice, hamsters, and gerbils contain approximately 24% crude protein and 4 % crude fiber.

An adult rat (300 gm) will consume approximately 5 gm feed and 10 ml water per 100 gm body weight per day. Consumption varies considerably with the ambient temperature and humidity, state of health, breeding status, and composition and quality of the diet.

BREEDING

Sexing a Rat

The testicles are evident at an early age, especially if the rat is held head up, and the testes pass from the inguinal canal into the scrotum. Males have a larger genital papilla and greater anogenital distance than females (5 mm in males and 2.5 mm in females at 7 days) (Figs. 28 and 29). Female nipples are visible between 8 and 15 days.

Breeding Program for Rats

Considerations in rat breeding programs include space available, strain fecundity, inbreeding systems, epide-

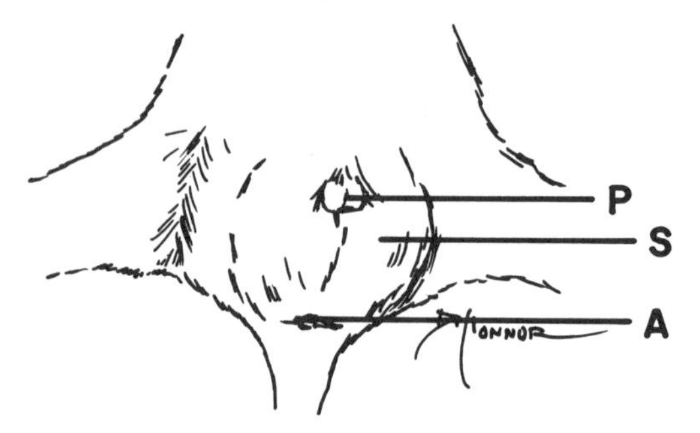

Fig. 28. External genitalia of the male rat. P = tip of penis; S = scrotal sac; A = anus.

miologic concerns, and production requirements. Rats are continuously polyestrus, with minor seasonal variations. Rats have a fertile, postpartum estrus occurring within 48 hours of parturition, but most rat breeders do not utilize the postpartum estrus. Simultaneous lactation and pregnancy may delay implantation 3 to 7 days.

Young rats are weaned at 21 days (40 to 50 gm). If the postpartum estrus is not utilized, the female rat resumes cycling 2 to 4 days postweaning.

Representative breeding systems for rats include monogamous and polygamous systems. The monogamous system involves the constant pairing of one male and one female, with the young removed prior to the succeeding parturition. This system requires a large male population and increased labor and equipment.

The polygamous or harem system combines 1 male and 2 to 6 females. Females are removed to separate cages prior to parturition; the postpartum estrus is not utilized. In this system each female provides more milk and larger young and litters. The female is returned to the harem for breeding after the young are weaned. If a colony system involves leaving the males and females together, removing the young for 12 hours on the first day postpartum may facilitate postpartum mating.

Some commercial rat producers move the male from cage to cage, each cage containing a single female. The male remains 1 week in each cage and is reintroduced just after weaning. One male can be utilized for every 7 females. If the male is not removed before parturition, cannibalism, litter desertion, and agalactia may result.

Strong, healthy, vigorous offspring are produced if the rats are first mated at 55 to 90 days of age, when the females weigh 250 gm and the males 300 gm. The age at first mating depends on the strain and growth rate of the rat.

Receptivity for Mating

The short estrous cycle reduces the necessity for cycle monitoring. During proestrus the vaginal smear contains epithelial cells with some cornified cells and leukocytes. In estrus, cornified cells become prominent. In late metestrus and during diestrus cornification decreases and lymphocytes increase.

Estrous Cycle in Rats

The rat estrous cycle lasts 4 to 5 days with an estrous period of approximately 12 hours, which, as in mice, occurs in the evening. The Whitten effect, a synchronization of estrus occurring after a group of females has been exposed to a male, is less pronounced in rats than in mice. Occasional induced ovulations are suspected in rats.

Gestation Period in Rats

Rats have a gestation period between 20 and 22 days long. Lactation delays

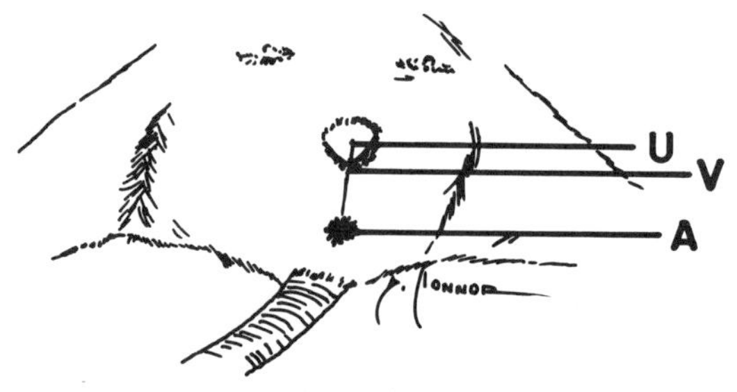

Fig. 29. External genitalia of the female rat. U = urethral orifice; V = vaginal orifice; A = anus.

implantation 5 to 7 days. The Bruce effect has not been reported in rats.

Determination of Pregnancy

After rats mate, a white, waxy copulatory plug is present in the vagina for 12 to 24 hours postcoitum. The plug may also be found in the cage or waste pan. If mating occurred, sperm will be present in the vaginal smear. Rats can be palpated, observed, or weighed for pregnancy detection. Mammary development is evident at 14 days.

Pseudopregnancy in Rats

Pseudopregnancy is uncommon in rats. When it occurs, it lasts approximately 13 days.

Nest Building

Rats build nests from bedding, but they require only small quantities for a satisfactory maternal nest. Cotton, tissue paper, soft wood shavings, and shredded newspaper make good nesting materials. Pups born on ground corn cob bedding are usually exposed to the plastic or metal cage floor.

Size of Litter

An average rat litter contains 6 to 12 young. Litter size and fertility decrease after eight to ten litters.

Abandonment and Cannibalism

For the first few days postpartum, and at weaning, disturbances such as excessive handling, loud noises, and lack of nesting material may cause the female rat to destroy the young. It may be necessary to isolate the mother before parturition. If the cage must be cleaned or changed, transfer some of the old bedding or the nest to the new cage.

Hand Rearing

Germ-free rats have been produced by hand raising under sterile techniques, and presumably young under 16 days could be fed a warmed formula with a pipette, but young rats are easily chilled, and the food may be aspirated. Simulated maternal stimulation of defecation and micturition must be provided.

DISEASE PREVENTION

In addition to the routine husbandry precautions listed in Chapter 1, a systematic diagnostic evaluation of colony animals should routinely be conducted to screen for subclinical infections. Filter cage covers prevent or reduce airborne transmission of microorganisms within a densely populated rat room, although such covers increase the ammonia level, humidity, and temperature within the cage.

PUBLIC HEALTH CONCERNS

Zoonotic diseases carried or transmitted by *Rattus norvegicus* include leptospirosis, *Streptococcus pneumoniae* infection, salmonellosis, *Hymenolepis* spp infection, and *Streptobacillus moniliformis* infection (rat bite fever in man). Sylvatic plague (*Yersinia pestis*) is carried by the common (wild) rat flea *Nosopsyllus fasciatus*. The mite *Liponyssus sylviarum* can transmit the virus of St. Louis encephalitis, and *Liponyssus bacoti* attacks man. These diseases and infections, with the exception of *Streptococcus* infections, are all extremely rare or unknown in domestic rats.

ANATOMIC AND PHYSIOLOGIC CHARACTERISTICS

Rats have no gall bladder, but the characteristics of the rat which most recommend its use in biomedical research are the unspecialized anatomy and physiology, the convenient size, and the well-defined behavior, anatomy, physiology, nutritional requirements, and responses to pathogens. Rats are commonly used in bioassay studies on nutritional and endocrinologic factors, in toxocologic studies, and in cancer research, immunology, and radiation biology.

Listings of available inbred rat strains can be obtained from the National Institutes of Health's *Catalogue of NIH Rodents,* NIH, Bethesda, MD 20014 (DHEW Publication NH-74-606) or from the Roscoe B. Jackson Memorial Laboratory, Bar Harbor, ME 04609.

BIBLIOGRAPHY

Bernstein, L.: A study of some enriching variables in a free-environment for rats. J. Psychosom. Res., *17*:85-88, 1973.

Blackmore, D. W.: Individual differences in critical temperatures among rats at various ages. J. Appl. Physiol., *29*:556-559, 1970.

Blackmore, D. W.: The effect of limited, moderate and lengthy daily separation from the mother during the early postnatal period of the rat on concurrent and subsequent growth, and on concurrent oxygen consumption at, as well as below, thermal neutrality. Biol. Neonate, *21*:268-281, 1972.

Farris, E. J., and Griffiths, J. Q.: The Rat in Laboratory Investigation. New York, Hafner, 1962.

Green, E. C.: Anatomy of the Rat. New York, Hafner, 1935.

Harris, J. M.: Differences in responses between rat strains and colonies. Toxicology, *3*:199-202, 1965.

Hughes, C. W., et al.: Domestication, sophistication, and avoidance in Norway rats. J. Comp. Physiol. Psychol., *84*:408-413, 1973.

Hughes, P. C., et al.: The effect of the number of animals per cage on the growth of the rat. Lab. Anim., *7*:293-296, 1973.

Lane-Petter, W., Lane-Petter, M. E., and Bowtell, C. W.: Intensive breeding of rats. 1. Crossfostering. Lab. Anim., *2*:35-39, 1968.

Lesser, G. T., et al.: Aging in the rat: longitudinal and cross-sectional studies of body composition. Am. J. Physiol., *225*:1472-1478, 1973.

Richie, D. H., and Humphrey, J. K.: Some observations on the mating of rats. J. Inst. Anim. Tech., *21*:100-105, 1970.

Weisbroth, S. H.: The Long-Evans rat in biomedical research. Lab. Anim. Care, *19*:697-755, 1970.

Chapter 3

Clinical Procedures

The clinical principles underlying the treatment of disease in individual laboratory animals are emphasized in this chapter; specific treatments are discussed with the several common diseases described in Chapter 5.

The chapter discusses drug dosing and dosages, anesthetics and anesthesia, surgical procedures, radiographic techniques, and euthanasia in laboratory animals.

Treatments and drug dosages for rabbits and rodents are often based on extrapolations from other species, limited clinical trials, and hearsay. In large colonies individual animals are seldom treated. Nevertheless, for the benefit of the pet owner and the researcher with the valuable subject, a treatment should at least be suggested or attempted, however poor the prognosis.

Drug Dosages and Therapeutic Regimens

Drug dosages and therapeutic regimens in rabbits and laboratory rodents are based, for the most part, on scattered references and observations. Few drugs commonly used in practice are specifically designated for use in laboratory animals, a consideration which presents both legal and therapeutic complications.

Apart from the shortcomings resulting from interspecies dosage extrapolation and limited clinical trials, the heterogeneous character of individual animals within a species is a major concern in dosage determination. Some of the several factors influencing drug effects include the species, breed, sex, age, diet, health status, biologic cycle phase, breeding status, metabolic rate, and nutritional level of the subject, as well as the composition of the bedding, the experimental protocol, use of other drugs, type of caging, and the ambient temperature. The dosages recommended in this section are approximate for the average population for the indications stated. Individual responses may vary, and animals receiving these levels may be unaffected, respond as desired, or be fatally overdosed. Other important considerations in drug therapy include toxicity, effect on experimental procedures, and the possible suppression of an active clinical infection to the chronic, subclinical, or carrier state.

The drugs included do not represent all the drugs that are efficacious in laboratory animals, and the intention is not to promote one product over another. The data in the section were assembled from personal observations and communications, from numerous scientific publications, from C. D. Barnes and L. G. Eltherington's *Drug Dosage in Laboratory Animals, A Handbook* (Berkeley, University of California Press, 1973), and from I. S. Rossoff's *Handbook of Veterinary Drugs: A Compendium for Research and Clinical Use* (New York, Springer Publishing Company, 1974). Before specific drugs, dosages, and indications are listed, information regarding average body weights, dosing and injection procedures, and daily water and feed consumption are described.

Adult Weights:	Approximate Ranges
Rabbit	2000–7000 gms
Guinea Pig	750–1000 gms
Hamster	90–150 gms
Gerbil	70–120 gms
Mouse	20–40 gms
Rat	400–800 gms

DOSING AND INJECTION PROCEDURES

	Oral	SQ	IM	IV
Rabbit	Feed or flexible tube	Flank, nape of neck	Thigh, lumbar muscles	Marginal ear vein, cephalic vein
Guinea Pig	Feed, flexible tube, bulbed needle	Over back	Posterior thigh muscles	Ear vein, saphenous vein, dorsal penile vein
Hamster	Feed, bulbed needle	Nape of neck, abdominal skin	Posterior thigh muscles	Femoral or jugular vein
Gerbil Mouse Rat	Feed, bulbed needle	Nape of neck, abdominal skin	Posterior thigh muscles	Lateral tail vein

The stomach tube for a rabbit may be a No. 5, 8, or 10 French catheter (2 to 3.5 mm), depending on the size of the rabbit. The tube is passed through a hole in the side of a one-half inch wooden dowel placed as a bit in the rabbit's diastema. The dowel will prevent chewing on the tube. Bulbed needles for rodents are available from Popper and Sons, Inc., 300 Denton, North Hyde Park, NY 11040.

The food and water needs listed below are approximations for dosing purposes only; actual amounts consumed depend on the nature of the diet (dry, powder, pellets, greens), ambient temperature, ventilation, hair coat, breeding and health status, and competition for feed and water. Amounts may be triple or more during lactation.

APPROXIMATE FOOD AND WATER NEEDS OF ADULT LABORATORY ANIMALS

Species	Daily Feed Intake per 100 gm Body Weight	Daily Water Intake per 100 gm Body Weight
Rabbit	5 gm	10 ml
Guinea Pig	5 gm	10 ml
Hamster	10 gm	10 ml
Gerbil	10 gm	5 ml
Mouse	15 gm	15 ml
Rat	5 gm	10 ml

TABLE OF DRUG DOSAGES

Drug	Indication	Dosage	Route	Remarks
Anticholinergic				
Atropine sulfate	Anticholinergic; respiratory stimulant	0.1–3.0 mg/kg	SC	Many rabbits and rats possess a serum atropinesterase
Antimicrobials				
Cephaloridine injectable	Broad spectrum antibiotic	10–25 mg/kg daily for 5–7 days	IM	Nephrotoxic in rabbits at doses above 200 mg/kg
Chloramphenicol palmitate	Broad spectrum antibiotic	50 mg/kg daily for 5–7 days	Orally	May prolong barbiturate anesthesia
Chloramphenicol succinate	Broad spectrum antibiotic	30 mg/kg daily for 5–7 days	IM	
Dimetridazole powder	Prevention of non-specific enteropathies and hexamitiasis	0.025–0.1% in drinking water during weaning	Orally	Federal Drug Administration (FDA) approved for turkeys only
Furazolidone powder or liquid	Prevention of enzootic pasteurellosis in rabbits	Supplemented in feed or given at 5 mg/kg	Orally	Nephrotoxic; not FDA approved for use in rabbit feed
Griseofulvin tablet	Antifungal agent	Rabbits 25 mg/kg daily for 4 wks; Guinea pigs 75 mg/kg daily for 2 wks	Orally	Treatment recommended only if accompanied by good husbandry
Oxytetracycline powder	Growth promotion	Supplemented in feed at 10 gm/ton feed	Orally	FDA approved as rabbit feed additive
Penicillin G injectable	Broad spectrum antibiotic	40,000 units/kg daily for 5–7 days	IM	May cause fatal floral alteration in gut of guinea pigs and hamsters
Sulfamerazine soluble	Prevention of murine respiratory disease and enteritis	1 mg/4 gm feed or 0.02% in water	Orally	

TABLE OF DRUG DOSAGES (*continued*)

Drug	Indication	Dosage	Route	Remarks
Sulfaquinoxaline soluble	Prevention of coccidiosis and acute pasteurellosis in rabbits	Continuously at 0.025% for 30 days; treatment level 1% in water for 2 wk	Orally	FDA approved additive in rabbit feed; vitamin K antagonist
Tetracycline soluble	Prevention and treatment of murine chronic respiratory disease	3–5 mg/ml in drinking water	Orally	Therapeutic dosages of tetracyclines are high for rodents
Anthelmintics				
Niclosamide tablet	Anthelmintic and taeniacide	Single oral dose 100 mg/kg or 4 mg/gm feed	Orally	
Piperazine powder	Anthelmintic	Rabbits 200 mg/kg; Mature rodents 3 mg/ml drinking water; young rodents 2 mg/ml water	Orally	Piperazine taste in water may be disguised by molasses
Thiabendazole	Anthelmintic	100–200 mg/kg	Orally	
Injectable Anesthetics				
Fentanyl-Droperidol (0.4 mg/ml–20 mg/ml) (Innovar-Vet[1])	Neuroleptanalgesic for induction or anesthesia	Rodents 0.13 ml/kg; mouse 0.005 ml/kg	IM	Causes extensive tissue necrosis in guinea pigs
Ketamine HCl injectable	Dissociative anesthetic for restraint and minor surgery	Restraint at 20 mg/kg; anesthesia at 40–60 mg/kg	IM	Not effective anesthetic in small rodents; muscle relaxation improved if acepromazine is added but increased chance of cardiac and pulmonary depression
Pentobarbital sodium	Barbiturate anesthetic	Dose to effect; suggested IV dose 25–50 mg/kg; IP dose 30–80 mg/kg	IV, IP	Dosages higher in smaller rodents; male rats, female mice, and animals on pine or cedar bedding require higher dosages. Reduce dose when used after tranquilizer

Thiamylal or tiopental	Ultrashort acting barbiturate anesthetic	IV dose 25–50 mg/kg; IP 45–50 mg/kg; dose to effect	IV, IP	Causes tissue irritation if injected extravascularly
Xylazine sodium	Sedative and Analgesic	7 mg/kg (with ketamine)	IM	Dosage for rabbits
Insecticides				
Carbaryl powder	Ectoparasite control	5% dust undiluted or 1:1 with talcum powder	Topically	Young rodents are susceptible to insecticide poisoning
Dichlorvos pellets or resin strip	Ectoparasite control	Section of strip placed in cage 24 hrs twice, 1 wk apart, or 2 ml pellets for 2 wks	Cage or bedding	Organophosphates have anticholinesterase activity; may inhibit breeding or potentiate barbiturate anesthesia
Oxytocic				
Oxytocin injectable	Delayed parturition or dystocia	0.2–3.0 units/kg	SC, IM	Determine cervical dilation before administering
Poison Antagonists				
Atropine sulfate	Organophosphate overdose	10 mg/kg every 20 min	SC	
Menadione	Warfarin poisoning	1–10 mg/kg as necessary	IM	
Tranquilizers				
Acepromazine maleate	Tranquilizer; preanesthetic	0.5–1.0 mg/kg	IM	May precipitate epileptiform seizures in gerbils; may produce hypothermia
Chlorpromazine	Tranquilizer; preanesthetic	7.5 mg/kg 25 mg/kg	IV IM	May cause myositis in rabbits and hypothermia and teratogenic effects in rats and mice
Diazepam	Tranquilizer; preanesthetic	5–10 mg/kg	IM	

[1] Pitman-Moore, Inc., P.O. Box 344, Washington Crossing, NJ 08560

ANTIBIOTIC THERAPY IN LABORATORY ANIMALS

Antibiotic therapy should include considerations of bacterial sensitivity, cost of the treatment regimen, toxicity of the antibiotic to the animal, public health significance of the pathogen involved, and the experimental protocol of the research study. Penicillin, bacitracin, and several other antibiotics should not be given to guinea pigs or hamsters. These antibiotics apparently cause an alteration in the intestinal flora (suppression of Gram-positive organisms), which leads to diarrhea or death within 3 to 5 days. If these "toxic" antibiotics should be given to guinea pigs or hamsters, a combination of 5 mg neomycin and 3 mg polymixin B given orally twice a day for 5 days may counteract the effects of the contraindicated drugs. Streptomycin and procaine are toxic for mice and other small rodents. Guinea pigs and hamsters may be treated with cephaloridine, chloramphenicol, and neomycin.

Anesthetics and Anesthesia

This section summarizes the scattered and often outdated information on anesthesia in laboratory animals. Detailed discussions of anesthetics and anesthetic techniques may be found in the several excellent, comprehensive papers on veterinary anesthesia listed in the section bibliography.

Laboratory animal species, with their great intraspecies and interspecies biologic diversity, high metabolic rates, small sizes, sharp teeth, varied metabolic processes, drug sensitivities, disease susceptibilities, lack of superficial veins, and often inaccessible glottis, present several unique difficulties for the anesthetist. Except for the use of the marginal ear vein and cephalic vein in the rabbit and the lateral tail vein in the mouse and rat, intravenous injections are difficult. Rabbits and rats have a high incidence of respiratory disease, which reduces the functional gas exchange surface and blocks air passages. Rabbits and guinea pigs have large ceca, which act as reservoirs for anesthetics. The respiratory centers in rabbits and rodents are particularly sensitive to anesthetic-induced suppression; this sensitivity is a problem in rabbits and guinea pigs, which hold their breath on first exposure to the inhalant anesthetics and then take a deep, often fatal, gasp. Mice, especially male mice of certain strains, develop renal and hepatic necrosis following chloroform exposure, and rabbits and guinea pigs respond with a severe myositis to injections of chlorpromazine and fentanyl-droperidol (Innovar-Vet) respectively. All small laboratory rodents lose body heat rapidly when anesthetized. Effective anesthetic dosages, particularly with sodium pentobarbital, vary widely, and a tranquilizing dose in one animal may be a lethal dose in another. Cross-species extrapolations are risky.

The concerns regarding anesthetic dose and the age, sex, pregnancy status, health, obesity, metabolic rate, nature of bedding, exposure to other drugs, and dietary intake of the patient are as applicable to laboratory animals as to other species. Rabbits and guinea pigs should be fasted for 12 hours prior to anesthesia.

Inhalation anesthetics may be delivered on a gauze pad placed in a covered container or through a face mask attached to a closed or semiclosed anesthetic machine. Oxygen flow should range from 100 ml/min for the smaller rodents to 1 L/min for rabbits. The bibliographic entries contain several suggestions for anesthetic devices and procedures.

The depth of anesthesia in laboratory animals is, without practice, difficult to determine. The most accurate statement on the subject is that the patient should be constantly monitored for heart beat and respiratory function.

The indications of level of anesthesia in laboratory animals are similar to those in other species. Respiratory rate, relaxation of the abdominal musculature, color of the mucous membranes, and certain reflex responses are used to determine the level. In the rabbit, signs of surgical anesthesia include the movement of the membrana nictitans over approximately one third of the cornea, a stable respiratory rate of 18 to 24 respirations per minute, relaxation of the abdominal musculature, and loss of the mouth, conjunctival, and toe pinch reflexes. Protrusion of the eyeball, dilated pupils, and cyanosis indicate an anesthetic overdose. The anesthetized guinea pig may move its limbs and create the false impression of recovery, but additional anesthetic may kill the animal. The pedal reflex is not a reliable indicator of surgical anesthesia in the guinea pig.

In case of an anesthetic overdose, the experimenter should place an opened plastic syringe cover or cupped hand over the animal's nose and periodically blow into the conduit to inflate the lungs.

TRANQUILIZERS

Tranquilizers or ataractics are drugs that induce a behavioral change characterized by relaxation and anxiety reduction without analgesia, anesthesia, ataxia, or loss of consciousness. Tranquilizers may be used for restraint or as preanesthetics. Tranquilizers discussed here are acepromazine maleate and chlorpromazine hydrochloride.

Acepromazine Maleate

Acepromazine is a phenothiazine derivative tranquilizer frequently used in conjunction with ketamine in laboratory animals. Although acepromazine has no analgesic properties, it provides an antiemetic effect and facilitates skeletal muscle relaxation. A common dosing procedure is to inject 1 ml (10 mg) acepromazine into 10 ml (1000 mg) ketamine and dose according to ketamine requirements (20 to 50 mg/kg IM). Acepromazine may precipitate epileptiform seizures in susceptible gerbils and cause hypothermia in animals.

Chlorpromazine Hydrochloride

Chlorpromazine is a phenothiazine derivative tranquilizer infrequently used in laboratory animals to reduce the barbiturate dosage. This potentiating effect, common to tranquilizers, may reduce the barbiturate requirement as much as 50%; however, the combination of tranquilizer and barbiturate may cause a precipitous fall in blood pressure. As the anesthetic dose of sodium pentobarbital in rabbits and rodents varies widely and may be dangerously close to the level of respiratory depression, chlorpromazine may be indicated. Chlorpromazine is given intramuscularly 30 minutes prior to the anesthetic; however, intramuscular injections of chlorpromazine (25 mg/kg) have caused a severe myositis in rabbits. Chlorpromazine in laboratory animals has also been associated with hypothermia and hyperglycemia.

ANTICHOLINERGICS

Anticholinergics are drugs that reduce the stimulatory effects of the parasympathetic nervous system on glands and smooth muscle and inhibitory effects on the heart. Anticholinergics may be used to reduce the copious upper respiratory secretions and bradycardia associated with certain anesthetic agents. Atropine sulfate is the anticholinergic discussed here.

Atropine Sulfate

Atropine, a competitive parasympathetic blocking agent, reduces respiratory tract secretions, anesthetic-induced bradycardia, and gastrointestinal motility. Atropine is available at 0.5 mg/ml. The dosages used in laboratory animals (0.1 to 3.0 mg/kg) are higher than dosages commonly recommended for cats and dogs. The drug is administered 30 minutes prior to the anesthetic. Some rabbits (approximately 30%) and rats possess a serum atropinesterase, which reduces the effect of a given dose of atropine.

NEUROLEPTANALGESICS, SEDATIVES, AND HYPNOTICS

Neuroleptanalgesia is a state of central nervous system depression and analgesia produced by a combination of narcotic analgesic and a tranquilizer. Innovar-Vet is the neuroleptanalgesic discussed here. Sedatives and hypnotics induce mild to deep central nervous system depression (anesthesia) with variable effects on pain perception. Ketamine, sodium pentobarbital, and paraldehyde are discussed.

Fentanyl-Droperidol (Innovar Vet)

Innovar-Vet is a neuroleptanalgesic containing the potent narcotic analgesic fentanyl (0.4 mg/ml) and the butyrophenone derivative tranquilizer droperidol (20 mg/ml). Innovar-Vet provides good analgesia, muscle relaxation, moderate bradycardia, and a tendency toward hypotension. The bradycardia may be counteracted by atropine. Onset of anesthetic activity requires approximately 10 to 15 minutes, and the effect lasts 1 to 3 hours. Innovar-Vet injected into guinea pigs causes a severe inflammatory reaction in muscle and surrounding tissues. Naloxone as a narcotic antagonist is given at 0.04 mg/kg intramuscularly every 4 to 15 minutes as needed. Nalorphine is given at 0.2 to 2.0 mg/kg intravenously.

Ketamine Hydrochloride

Ketamine is a cyclohexamine derivative and dissociative anesthetic used in rabbits and rodents for restraint and minor surgical procedures and as an inducing agent before inhalation anesthesia. Acepromazine (1 mg) or xylazine (14 mg) combined with 100 mg ketamine improves skeletal muscle relaxation. Ketamine causes mild cardiovascular stimulation and moderate respiratory depression. Induction time following intramuscular injection is approximately 5 to 12 minutes and duration of anesthesia 15 to 30 minutes. Analgesia with ketamine may be more effective for somatic than visceral pain. Ketamine produces a quieting effect in rabbits and guinea pigs, with poor to good muscular relaxation, but the reported effects in smaller rodents vary from satisfactory anesthesia at 44 mg/kg to little effect at 100 mg/kg. Hamsters and mice injected intramuscularly with dosages above 100 mg/kg may continue to walk unsteadily around the cage. Ketamine hydrochloride injectable is available in a 100 mg/ml solution.

Pentobarbital Sodium

Sodium pentobarbital is an injectable barbiturate anesthetic commonly used intravenously or intraperitoneally as an anesthetic in rabbits and rodents. The dosage necessary for anesthesia is variable and often close to the lethal dose; therefore, a given amount of drug may have little effect on one animal and kill another. As sodium pentobarbital should always be administered slowly to effect, many investigators suggest diluting the 6% solution with physiologic saline solution. Induction via the intraperitoneal route requires 5 to 15 minutes, and anesthesia lasts approximately 30 to 45 minutes. Recovery is prolonged. The use of chlorpromazine intravenously at 7.5 mg/kg reduces the anesthetic dosage of sodium pentobarbital by approximately half. Male rats, female mice, and rodents housed on pine or cedar bedding require somewhat higher doses for anesthesia. Sleeping time is extended in animals exposed to organophosphate insecticides. Sodium pentobarbital provides good analgesia and muscle relaxation.

Paraldehyde

Paraldehyde, a colorless liquid with a disagreeable odor, decomposes or is oxidized readily into acetaldehyde and acetic acid. It is a rapidly acting hypnotic with no analgesic properties and has been used in conjunction with ketamine in rabbits at 0.5 mg/kg.

INHALANT ANESTHETICS

Inhalant anesthetics are volatile gases that produce loss of consciousness, analgesia, and varying degrees of skeletal muscle relaxation. Diethyl ether, halothane, and methoxyflurane are discussed.

Diethyl Ether

Ether is a highly soluble, volatile, and flammable inhalation anesthetic with excellent analgesic and muscle relaxation properties. Ether, however, is an irritating gas, and premedication with atropine is necessary to prevent copious production of respiratory secretions. Breath holding followed by an often fatal gasp occurs with ether anesthesia in guinea pigs and rabbits. Induction to surgical anesthesia requires 1 to 15 minutes, depending on the species and the ether concentration involved.

Halothane

Halothane is a volatile, nonflammable, halogenated ethane that can be used alone or following ketamine induction to produce surgical anesthesia. Rapid induction, absence of effect on respiratory secretions, and ease of control are advantages of halothane. Disadvantages include severe cardiovascular depression, sensitization of the myocardium to epinephrine, and moderate analgesic and muscle relaxation properties. Halothane may be administered in a closed container or via a face mask and closed or semiclosed anesthetic system. Rabbits may be intubated, but mastering the technique requires practice, as the mouth opening is small, the teeth are sharp, and the glottis is located deep in the oral cavity. Halothane gas concentration, for induction and maintenance, is 0.1% to 2.5% with an oxygen flow 100 ml to 1 L per minute. Induction concentrations should begin at the lower levels and be increased as required. Nitrous oxide facilitates induction and allows for maintenance at lower concentrations of halothane. The bradycardia effect may be reduced with atropine premedication.

Methoxyflurane

Methoxyflurane is a nonflammable, halogenated ether of low volatility. As mild depressive effects on respiratory and cardiovascular function may occur with methoxyflurane, premedication with atropine is variably advised. Although induction is slow (10 to 15 min), muscle relaxation and analgesia are excellent. The induction may be eased and shortened with the use of acepromazine-ketamine as an inducing agent and nitrous oxide combined 1:1 with oxygen. Up to a 3% gas concentration is used for induction and a 0.1% to 1.5% concentration for maintenance.

ANESTHESIA IN THE RABBIT

In small rodents and rabbits evidence of surgical anesthesia is based on loss of certain reflexes, a stable respiratory rate, and on the relaxation of the abdominal musculature. The rabbit has a respiratory rate between 35 and 60 respirations per minute and a tidal volume of 2.25 ml/kg body weight. Ether and sodium pentobarbital (35 to 55 mg/kg) are "high risk" anesthetics in rabbits because of the sensitivity of the rabbit's respiratory center to volatile anesthetics and injected barbiturates. Ketamine alone at 25 mg/kg will quiet a rabbit for examination or intubation. Because of the serum enzyme atropinesterase, which hydrolyzes atropine, rabbits have a variable response to a given dose of the drug. A suggested dose for atropine in laboratory animals is between 0.1 and 3.0 mg/kg body weight.

Intubation in the rabbit is difficult, but with proper equipment and restraint the technique can be mastered. A soft wire stylet is placed inside a 3 to 4 mm endotracheal tube and bent to form a three-eighth circle corresponding to the length of the rabbit's mandible. A laryngoscope with a blade 75 to 158 mm long facilitates passage of the tube over the extended tongue and into the glottis. Topical anesthetics to prevent laryngospasm are not indicated because they suppress the convenient forward motion of the glottis during swallowing and may initiate coughing. Respiratory disease, which is common in rabbits, could cause reduced surface for gas exchange and obstructed air passages.

Ketamine alone or in combination with 1 mg acepromazine or 14 mg xylazine/100 mg ketamine is given intramuscularly at 25 mg/kg for tranquilization and 40 mg/kg for anesthesia. This injection may be followed as needed by a second ketamine injection (20 mg/kg) or by methoxyflurane or halothane to the desired level of anesthesia. The inhalation anesthetic may be administered by nose cone or endotracheal tube. Ketamine alone provides poor analgesia and skeletal muscle relaxation in rabbits.

Ketamine (50 mg/kg) and paraldehyde (0.5 mg/kg) can be administered intramuscularly. After 2 hours ketamine may be given again at 20 mg/kg to maintain anesthesia.

Methoxyflurane or halothane may be administered alone or with nitrous oxide until surgical anesthesia is reached, at which time the gas concentration is reduced for proper maintenance. Rabbits exposed to halothane often struggle and cough.

Innovar-Vet (fentanyl-droperidol) is administered intramuscularly at 0.13 to 0.17 ml/kg body weight. Sodium pentobarbital is given slowly via the marginal ear vein until the desired level of anesthesia is achieved. The dose is between 35 and 55 mg/kg, although respiratory depression is present at dosages of 50 mg/kg. The higher doses can be reduced to 20 or 30 mg/kg if the rabbit is given chlorpromazine at 7.5 mg/kg intravenously or 25 mg/kg intramuscularly.

Hypnosis has been used as a restraining technique for rabbits for injections, radiography, and even castrations. The rabbit is firmly restrained on its back, with the eyes covered or uncovered, in a V-shaped rack. The ventrum is gently stroked until the rabbit relaxes and breathing slows from the normal average of 51 to 18 to 24 respirations per minute. If the rabbit awakens during the procedure, restrain the animal and resume stroking. Avoid sudden noises. Some investigators claim that tranquilizers, soft music, and gentle words facilitate hypnosis.

ANESTHESIA IN THE GUINEA PIG

Highly volatile inhalation anesthetics (ether, halothane) are "high risk" anesthetics in guinea pigs because the guinea pig (or rabbit) exposed to an irritating gas will first hold its breath and then breathe deeply. Halothane (1%) causes a 35% to 40% decrease in blood pressure in the guinea pig. Innovar-Vet (fentanyl-droperidol) injected intramuscularly in guinea pigs produces nerve, vessel, and muscle necrosis and sloughing of extremities. A guinea pig has a normal respiratory rate between 70 and 130 respirations per minute and an approximate tidal volume of 3.8 ml/kg body weight.

Acepromazine (1 ml or 10 mg) is injected into 10 ml (1000 mg) ketamine. This acepromazine-ketamine mixture is injected intramuscularly at a ketamine dose of 20 to 40 mg/kg body weight. The injected drugs are followed by methoxyflurane via nose cone. An adequate oxygen supply must be provided. At 22 mg/kg IM ketamine will quiet a guinea pig for approximately 15 minutes, although analgesia is poor. At 33 mg/kg the anesthesia lasts 26 minutes and at 44 mg/kg 35 minutes, but surgical anesthesia is not attained at the higher dose. Recovery requires 5 to 30 minutes.

Methoxyflurane is an excellent anesthetic for use in guinea pigs. Induction is smooth and muscle relaxation is excellent, but control of the depth of anesthesia is difficult if the open drop or nose cone method is used. Atropine should be used as a preanesthetic to prevent excessive salivation. If a closed or semiclosed anesthetic machine is used, 1% methoxyflurane is used for induction and 0.3% for maintenance. Induction requires 15 to 18 minutes and recovery 10 minutes or more.

Sodium pentobarbital administered intraperitoneally is difficult to dose because of the indefinite injection site, differential absorption, and varying dispersion of the drug in the peritoneal cavity. Sodium pentobarbital may be given intraperitoneally at 28 to 35 mg/kg body weight. This dose will provide approximately 30 minutes of surgical anesthesia. A combination of chlorpromazine at 25 mg/kg and sodium pentobarbital at 30 mg/kg will extend the anesthesia to 50 or more minutes. Guinea pigs should be fasted 12 hours prior to barbiturate anesthesia to prevent vomiting and dose miscalculation.

ANESTHESIA IN THE HAMSTER

Hamsters have a respiratory rate of approximately 40 to 120 respirations per minute and a tidal volume of 0.8 ml. The effect of ketamine in hamsters is variable; some investigators report anesthesia at 20 to 30 mg/kg, but others note only catalepsy and mild analgesia at 200 mg/kg body weight.

Sodium pentobarbital administered intraperitoneally at 60 to 90 mg/kg produces anesthesia lasting 30 to 45 minutes in hamsters. The injection is given in the lower abdomen off the midline.

Volatile anesthetics (ether, methoxyflurane) placed on a gauze sponge in a covered container will anesthetize hamsters for a short period. Innovar-Vet may be injected intramuscularly at 0.15 ml/100 gm body weight.

ANESTHESIA IN THE GERBIL

Sodium pentobarbital administered intraperitoneally at 0.01 ml/10 gm body weight to a maximum dose of 0.10 ml (6 mg) will anesthetize a gerbil for 30 to 45 minutes.

Ketamine (40 mg/kg) intramuscularly plus methoxyflurane via a nose cone (paper cup or plastic syringe cover) will produce analgesia in 90 seconds and deep analgesia in approximately 12 minutes.

Volatile anesthetics may be used in a covered container to anesthetize gerbils for short periods.

ANESTHESIA IN THE MOUSE

The mouse has an average respiratory rate of 60 to 220 respirations per minute and an approximate tidal volume of 0.15 ml. Intravenous injections are easily made into the lateral tail vein if the tail is pulled through a hole in the cage or in a piece of cardboard. Warming the tail dilates the vessels and facilitates entry.

Care should be taken not to aspirate blood or inject air emboli.

Sodium pentobarbital is a satisfactory anesthetic in the mouse when injected intraperitoneally at 80 to 90 mg/kg body weight. The pentobarbital solution should be diluted 1:9 with physiological saline solution. A combination of chlorpromazine intramuscularly (50 mg/kg) and sodium pentobarbital intraperitoneally (50 mg/kg) produces anesthesia in mice.

Ether, halothane, and methoxyflurane may be applied to a gauze and placed in a closed container. Ether causes profuse salivation in mice.

Although ketamine has been reported to produce anesthesia at 44 mg/kg body weight, other trials have reported that doses to 200 mg/kg are ineffective.

ANESTHESIA IN THE RAT

Rats have an approximate respiratory rate of 65 to 110 respirations per minute and an approximate tidal volume of 1.0 ml. Rats may be rolled in a small cloth to facilitate restraint. Rats housed on cedar bedding have reduced barbiturate sleeping time compared to rats housed on other bedding; this phenomenon may be due to the induction of certain hepatic enzymes by a component of the cedar bedding. Respiratory disease is common in rats.

Sodium pentobarbital administered intraperitoneally at 10 to 30 mg/kg in young rats and 30 to 50 mg/kg in older rats (male rats require the higher doses) produces light anesthesia at lower doses and deep anesthesia at the higher levels.

Ketamine hydrochloride at 10 to 60 mg/kg body weight produces poor relaxation and analgesia, but if ketamine is given at 60 mg/kg and sodium pentobarbital intraperitoneally at 20 mg/kg, anesthesia will last approximately 1 hour.

Innovar-Vet given intramuscularly 0.02 mg/100 gm will anesthetize a rat.

Surgical Techniques

Descriptions of surgical procedures in laboratory animals are frequently encountered in the literature, but usually the surgery itself is a means to an end in an experiment rather than a method to remedy an abnormality. This section summarizes surgical techniques that have been gleaned from the research literatue on rabbits and rodents.

During the surgical procedure a heating pad placed beneath the animal will help to maintain normal body temperature (small rodents, 37°-38° C; rabbits and guinea pigs, 39° C). With the heating pad, however, care must be taken to prevent overheating the patient.

Important postoperative considerations include a clean, quiet, warm environment (32° C or 85° F) and fluids (warmed 5% glucose) administered subcutaneously.

If the tail of a rabbit or rodent is grasped distal to the base, the tail skin may be pulled off. The skinless tail should be amputated, and the skin should be sutured over the stump.

Fracture reduction and fixation present no unique problems in laboratory animals except for the small size of the tissues involved and the low density of the long bones. However, the animal's movements and chewing behavior usually cause fixation devices to break down. Chlorpromazine may be used as a tranquilizer (2 to 25 mg/kg) to reduce splint chewing and suture removal. Rolled X-ray film may be used as a splint, or the fracture simply may be reduced and the animal given cage rest.

Castration of rabbits and guinea pigs is complicated by the open inguinal canal, which predisposes to intestinal herniation. The cutaneous incisions may be ample (2 to 3 cm), but the tunica vaginalis should be opened distally only enough to permit removal of the testes. Simple ligation of the vessels and ducts leading to or from the testes will effectively castrate a guinea pig. During castration procedures the testes must be manipulated from the inguinal canal into the scrotal pouches.

Ovariectomy of guinea pigs involves bilateral, retroperitoneal incisions made immediately ventral to the erector spinae muscles and 1 cm posterior to the last rib. As guinea pigs may have ovarian rests on the oviduct, cycling may continue despite the ovariectomy.

Hysterectomy in the rabbit involves tying the animal on a board tipped "head down" to displace the viscera craniad. A 5-cm midline incision beginning just cranial to the pubic symphysis exposes the peritoneum. The uterine horns, lying on either side of the bladder, are retracted and removed with the ovaries. Ligations are performed as in other species. Tissues and vessels in rabbits tear easily. The rabbit's cervix is divided.

In the rabbit the mediastinum can be entered through sternal splitting without causing a pneumothorax.

For orchidectomy in the mouse, the peritoneum is entered via a small, transverse, suprapubic incision. The vas deferens and vessels are transected, and the testes, epididymis, fat, and gubernaculum are removed. Hemostasis is rarely necessary. The cutaneous wound is closed with 3-0 silk sutures.

Oophorectomy in the mouse requires a dorsal, transverse, cutaneous incision at the second lumbar level. The skin is moved to reveal the periovarian fat visible through the body wall. The wall is entered, and the fat, ovary, and distal uterus are removed. The skin is closed with 1 or 2 silk sutures.

Although blood withdrawal is not a common diagnostic procedure, many experiments require the collection of blood from rabbits and rodents. All animals may be bled from the heart, either by inserting the needle directly over the heart or passing anterior alongside the xiphoid process. If blood collects in the pericardial sac, cardiac tamponade and death may result. Rats, mice, and gerbils can be bled from the lateral tail vein if the vein is first warmed in water or under a light. Rabbits and guinea pigs may be bled from the ear vein by needle or incision. A toe clip in a mouse or hamster will furnish a few drops of blood. The jugular, femoral, or cephalic veins can also be used. In gerbils, mice, and hamsters the retroorbital sinus may be entered by passing a polished microhematocrit tube alongside the eyeball at the medial canthus. The venous sinus is pierced, and the blood flows through the tube into a collection device. Although this technique appears stressful and damaging, healing is complete, and the animal may be rebled within a few days. Injection and withdrawal procedures are facilitated if the animal is tranquilized, anesthetized, or properly restrained.

Radiography

Radiography in laboratory animals may contribute to both experimental and diagnostic efforts. Abnormal conditions that may require radiography include arthritis, scorbutus, malocclusion, luxation or fracture of a limb or the spine, congenital abnormalities, otitis media, pyometra, neoplasia, abscessation, and gastric hairballs. Visualization of a gastric hairball is facilitated if a few milliliters of air are passed into the stomach.

Animals may be restrained for radiographic examination by manual means, with tranquilizers or anesthetics, or by hypnosis. The choice of kilovoltage depends on the tissue to be examined, the films and screens used, and the techniques employed. Higher kilovolt ranges permit reduction of exposure time and allow for more exposure latitude (more shades of grey) and organ observation; however, fine bone detail is lost. Lower kilovolt levels, requiring longer exposure times, will reveal fine bone detail, but the increased contrast reduces definition of soft tissues. Milliamperage can be manipulated somewhat to compensate for lower kilovoltages or greater subject-tube distances. Detailed films and screens should be used.

Comprehensive discussions of radiologic techniques in research animals are contained in H. G. Matthew and H. J. Barnard's chapter "Radiographic Techniques" in *Roentgen Techniques in Laboratory Animals*, B. Felson, Editor (Philadelphia, W. B. Saunders Company, 1968) and in W. B. Carlson's chapter "Radiography" in *Methods of Animal Experimentation*, Volume I, W. Guy, Editor (New York, Academic Press, Inc., 1965).

Euthanasia

Euthanasia is the humane killing of an animal. Implied in humane killing is a painless, quick, quiet transition from consciousness to unconsciousness and death. The euthanasia procedure should be carried out by a trained technologist using methods, substances, and facilities permitted under institutional policies and governmental laws. Animals should not be killed while crowded together or in the presence of living animals. The method of euthanasia varies with the species and intended postmortem use of the animal.

A common method for euthanasia involves the intracardiac, intravenous, or intraperitoneal injection of an overdose of sodium pentobarbital. Concentrated barbiturate solutions specifically intended for euthanasia are available. Barbiturates are controlled substances, however, and are not available to all animal care personnel.

The inhalant gases carbon dioxide and halothane are commonly used for humane killing. Liquid halothane or ether should not come into contact with the animals. Chloroform is toxic for mice and should not be used in mouse housing areas. Ether is irritating to the eyes and respiratory tract. Hot exhaust gases are not permitted as euthanasia agents.

Physical methods, such as cervical or cranial fracture, are satisfactory for euthanasia if conducted by experienced personnel on small numbers of animals.

Serologic Testing for Rodent Viruses

Viral infections in rodents, although often latent, are common and may influence experimental results or predispose the infected animal to clinical viral or bacterial disease. Ectromelia virus (mousepox) can cause devastating epizootics in mouse colonies, and Sendai virus infection is one of the most common and serious diseases among mice in the United States. The detection of rodents carrying antiviral antibodies, and possibly the virus itself, is therefore an important consideration in disease prevention in research animal colonies. The chart below summarizes the more common serologic tests suggested for screening rodent sera. Hemagglutination inhibition (HI) and complement fixation (CF) tests are included.

The collection, processing, dilution, and storage of rodent sera intended for serologic testing must follow an established protocol if the results are to be meaningful. Collection tubes must be clean and held in an ice bath. Following collection of the blood, the clotted sample is centrifuged, the serum is removed and diluted 1:5, then heat inactivated prior to testing. Information on serum processing, available diagnostic services, and prices are available from Microbiological Associates, Viral Testing Laboratory, 4733 Bethesda Avenue, Bethesda, MD 20014. Information on mousepox may be obtained from the

COMMON SEROLOGIC TESTS

	Guinea Pig	Hamster	Mouse	Rat
Reo 3 (HI)	x	x	x	x
PVM (HI)	x	x	x	x
Kilham rat (HI)				x
K virus (HI)			x	
GD VII (HI)	x	x	x	x
Sendai (HI)	x	x	x	x
Ectromelia (HI)			x	
Mouse adenovirus (CF)			x	x
MHV (CF)			x	x
LCM (CF)	x	x	x	x

Microbiological Associates at the address given above.

BIBLIOGRAPHY

Drug Dosages and Therapeutic Regimens

Baer, H.: Long-term isolation stress and its effects on drug response in rodents. Lab. Anim. Sci., *21*:341-349, 1971.

Bolton, G. R.: Aerosol therapy. *In* Current Veterinary Therapy V. Edited by R. W. Kirk. Philadelphia, W. B. Saunders, 1974.

Finco, D. R.: Fluid therapy. *In* Current Veterinary Therapy V. Edited by R. W. Kirk, Philadelphia, W. B. Saunders, 1974.

Harrison, J. B., Sussman, H. H., and Pickering, D. E.: Fluid and electrolyte therapy in small animals. J. Am. Vet. Med. Assoc., *137*:637-645, 1960.

Miller, G. H.: Aerosol therapy in acute and chronic respiratory disease. Arch. Intern. Med., *131*:148-155, 1973.

Milner, N. A.: Letter: penicillin toxicity in guinea pigs. Vet. Rec., *96*:554, 1975.

Morgan, D. R.: Routine birth induction in rabbits using oxytocin. Lab. Anim., *8*:127-130, 1974.

Muller, G. H.: Topical dermatologic therapy. *In* Current Veterinary Therapy V. Edited by R. W. Kirk. Philadelphia, W. B. Saunders, 1974.

Owens, D. R., Wagner, J. E., and Addison, J. B.: Antibiograms of pathogenic bacteria isolated from laboratory animals. J. Am. Vet. Med. Assoc., *167*:605-609, 1975.

Shah, D. V., and Suttie, J. W.: Vitamin K requirement and warfarin tolerance in the hamster. Proc. Soc. Exp. Biol. Med., *150*:126-128, 1975.

Woodard, G.: Principles in drug administration. *In* Methods of Animal Experimentation. Vol. 1. Edited W. I. Gay. New York, Academic Press, 1965.

Anesthetics and Anesthesia

Ben, M., Dixon, R. C., and Adamson, R. H.: Anesthesia in the rat. Fed. Proc., *28*:1522-1527, 1969.

Bree, M. M., Cohen, B. J., and Abrams, G. D.: Injection lesions following intramuscular administration of chlorpromazine in rabbits. J. Am. Vet. Med. Assoc., *159*:1598-1602, 1971.

Carvell, J. E., and Stoward, P. J.: Halothane anaesthesia of normal and dystrophic hamsters. Lab. Anim., *9*:345-352, 1975.

Chaffee, V., and Parkash, V.: A satisfactory method of anesthetizing rabbits for major or minor surgery. J. Am. Vet. Med. Assoc., *163*:664, 1973.

Clifford, D.: Restraint and anesthesia of small laboratory animals. *In* Textbook of Veterinary Anesthesia. Edited by L. R. Soma. Baltimore, Williams & Wilkins, 1971.

Collins, T. B., and Lott, D. F.: Stock and sex specificity in the response of rats to pentobarbital sodium. Lab. Anim. Care, *18*:192-194, 1968.

Connell, H.: Pentobarbitone sodium anesthesia for oral and immunological procedures in the guinea pig. Lab. Anim., *6*:55-60, 1972.

Dahloef, L. G., van Dis, H., and Larsson, K.: A simple device for inhalational anesthesia in restrained rats. Physiol. Behav., *5*:1211, 1970.

Davis, N. L., and Malinin, T. I.: Rabbit intubation and halothane anesthesia. Lab. Anim. Sci., *24*:617-621, 1974.

Dobson, C., and Tschirky, H.: Development of an anesthetic apparatus for experimental surgery on rats. Pharmacology, *5*:307-313, 1971.

Dudley, W. R., et al.: An apparatus for anesthetizing small laboratory animals. Lab. Anim. Sci., *25*:481-482, 1975.

Dulowy, W. C., Mombelloni, P., and Hesse, A. L.: Chlorpromazine premedication with pentobarbital anesthesia in the mouse. Am. J. Vet. Res., *21*:156, 1960.

Freeman, M. J., Bailey, S. P., and Hodesson, S: Premedication, tracheal intubation, and methoxyflurane anesthesia in the rabbit. Lab. Anim. Sci., *22*:576-580, 1972.

Green, C. J.: Neuroleptanalgesic drug combination in the anesthetic management of small laboratory animals. Lab. Anim., *9*:161-168, 1975.

Hoar, R. M.: Anesthesia in the guinea pig. Fed. Proc., *28*:1517-1521, 1969.

Hoar, R. M.: The use of metofane (methoxyflurane) anesthesia in guinea pigs. Allied Vet., *36*:131-134, 1964.

Hoar, R. M.: Anesthetic techniques for the rat and guinea pig. J. Am. Pharmacol. Assoc., *29*:325-343, 1940.

Hoge, R. S., et al.: Intubation technique and methoxyflurane administration in rabbits. Lab. Anim. Care, *19*:593-595, 1969.

Holland, A. J. C.: Laboratory animal anesthesia. Can. Anaesth. Soc. J., *20*:693-705, 1973.

Kent, G. M.: General anesthesia in rabbits using methoxyflurane, nitrous oxide, and oxygen. Lab. Anim. Sci., *21*:256-257, 1971.

Kinsey, V. E.: The use of sodium pentobarbital for repeated anesthesia in the guinea pig. J. Am. Pharmacol. Assoc., *29*:342-346, 1940.

Kisloff, B.: Ketamine-paraldehyde anesthesia for rabbits. Am. J. Vet. Res., *36*:1033-1034, 1975.

Lamkin, R. H., and McPherson, D. L.: Inhalation anesthesia for the short-term guinea pig experiment. Arch. Otolaryngol., *101*:138-139, 1975.

Leash, A. M., Beyer, R. D., and Wilber, R. G.: Self-mutilation following Innovar-Vet injections in the guinea pig. Lab. Anim. Sci., *23*:720-721, 1973.

Lewis, G. E., and Jennings, P. B.: Effective sedation of laboratory animals using Innovar-Vet. Lab. Anim. Sci., *22*:430-432, 1972.

Lindquist, P. A.: Induction of methoxyflurane anesthesia in the rabbit after ketamine hydrochloride and endotracheal intubation. Lab. Anim. Sci., *22*:898-899, 1972.

McCormick, M. J., and Ashworth, M. A.: Acepromazine and methoxyflurane anesthesia of immature New Zealand White rabbits. Lab. Anim. Sci., *21*:220-223, 1971.

McIntyre, J. W. R.: An introduction to general anesthesia of experimental animals. Lab. Anim., *5*:99-114, 1971.

Mauderly, J. L.: An anesthetic system for small laboratory animals. Lab. Anim. Sci., *25*:331-333, 1975.

Medd, R. K., and Heywood, R.: A technique for intubation and repeated short-duration anesthesia in the rat. Lab. Anim. *4*:75-78, 1970.

Morgan, W. A., et al.: Pentobarbital anesthesia in the rabbit. Am. J. Vet. Res., *27*:1133-1134, 1966.

Murdock, H. R.: Anesthesia in the rabbit. Fed. Proc., *28*:1510-1516, 1969.

Newton, W. M., Cusick, P. K., and Roffe, M. R.: Innovar-Vet induced pathologic changes in the guinea pig. Lab. Anim. Sci., *25*:597-601, 1975.

Paterson, R. C., and Rowe, A. H. R.: Surgical anesthesia in conventional and gnotobiotic rats. Lab. Anim., *6*:147-154, 1972.

Reves, J. G., and McCracken, Jr., L. E.: Halothane in the guinea pig. Anesthesiology, *42*:230-231, 1975.

Rubbright, W. C., and Thayer, C. B.: The use of Innovar-Vet as a surgical anesthetic for the guinea pig. Lab. Anim. Care, *20*:989-991, 1970.

Sawyer, D. C.: Anesthetic techniques of rabbits and mice. *In* Experimental Animal Anesthesiology. Edited by D. C. Sawyer. Brooks Air Force Base, Texas, 1965.

Schaffer, A.: Anesthesia and sedation. *In* Methods of Animal Experimentation. Vol. 1. Edited by W. I. Gay. New York, Academic Press, 1965.

Skartvedt, S. M., and Lyon, N. C.: A simple apparatus for inducing and maintaining halothane anesthesia of the rabbit. Lab. Anim. Sci., *22*:922-924, 1972.

Smith, S. M., and Kaplan, H. M.: Ketamine-methoxyflurane anesthesia for the Mongolian gerbil, *Meriones unguiculatus*. Lab. Anim., *8*:213-216, 1974.

Strack, L. E., and Kaplan, H. M.: Fentanyl and droperidol for surgical anesthesia of rabbits. J. Am. Vet. Med. Assoc., *153*:822-825, 1968.

Stunkard, J. A., and Miller, J. C.: An outline guide to general anesthesia in exotic species. Vet. Med. Small Anim. Clin., *6*:1181-1186, 1974.

Taber, R., and Irwin, S.: Anesthesia in the mouse. Fed. Proc., *28*:1528-1532, 1969.

Valenstein, E. S.: A note on anesthetizing rats and guinea pigs. J. Exp. Anal. Behav., *4*:6, 1961.

Ward, G. S., Johnsen, D. O., and Roberts, C. R.: The use of CI 744 as an anesthetic for laboratory animals. Lab. Anim. Sci., *24*:737-742, 1974.

Wass, J. A., Keene, J. R., and Kaplan, H. M.: Ketamine-methoxyflurane anesthesia for rabbits. Am. J. Vet. Res., *35*:317-318, 1974.

Weisbroth, S. H., and Fuden, J. H.: Use of ketamine HCl as an anesthetic in laboratory rabbits, rats, mice, and guinea pigs. Lab. Anim. Sci., *22*:904-906, 1972.

Willoughby, C. P.: Anesthetizing hamsters and small rodents. Vet. Rec., *92*:572-573, 1973.

Youth, R. A., et al.: Ketamine anesthesia for rats. Physiol. Behav. *10*:633-636, 1973.

Hypnosis

Gruber, R. P., and Amato, J. J.: Hypnosis for rabbit surgery. Lab. Anim. Care, *20*:741-742, 1970.

Klemm, W. R.: Drug potentiation of hypnotic restraint of rabbits as indicated by behavior and brain electrical activity. Lab. Anim. Care, *15*:163-167, 1965.

Rapson, W. S., and Jones, T. C.: Restraint of rabbits by hypnosis. Lab. Anim. Care, *14*:131-133, 1964.

Surgical Techniques

Anderson, M., and Froimovitch, M.: Simplified method of guinea pig castration. Can. Vet. J., *15*:126-127, 1974.

Biven, W. S., and Timmons, E. H.: Basic biomethodology. *In* The Biology of the Laboratory Rabbit. Edited by S. H. Weisbroth, R. E. Flatt, and A. L. Kraus. New York, Academic Press, 1974.

Brocklehurst, W. E.: An injection block for small animals. Lab. Pract., *21*:731, 1972.

Castro, J. E.: Surgical procedures in small laboratory animals. J. Immunol. Methods, *4*:213-216, 1974.

Chaffee, V. W.: Surgery of laboratory animals. *In* Handbook of Laboratory Animal Science. Edited by E. C. Melby and N. H. Altman. Cleveland, Chemical Rubber Company Press, 1974.

Di Pasquale, G., and Campbell, W. A.: A gag for gastric intubation of rabbits. Lab. Anim. Care, *16*:294-295, 1966.

Fraser, T., and Ascoli, R. C.: The castration of guinea pigs. J. Inst. Anim. Tech., *21*:21-24, 1970.

Furner, R. L., and Mellett, L. B.: Mouse restraining chamber for tail vein injection. Lab. Anim. Sci., *25*:648-649, 1975.

Hoar, R. M.: Methodology. *In* The Biology of the Guinea Pig. Edited by J. E. Wagner and P. J. Manning. New York, Academic Press, 1976.

Hurwitz, A. J.: A simple method for obtaining blood samples from rats. J. Lab. Clin. Med., *78*:172-174, 1971.

Moreland, A. F.: Collection and withdrawal of body fluids and infusion techniques. *In* Methods of Animal Experimentation. Vol. 1. New York, Academic Press, 1965.

Radiography

Berg, N. O., Idbohrn, H., and Wendeberg, B.: Investigation of the tolerance of the rabbit's kidney to newer contrast media in renal angiography. Acta Radiol., *50*:285-292, 1958.

Carlson, W. D.: Radiography. *In* Methods of Animal Experimentation. Vol. 1. Edited by W. I. Gay. New York, Academic Press, 1965.

Engeset, A.: Intralymphatic injections in the rat. Cancer Res., *19*:277-278, 1959.

Felson, B., (ed.): Roentgen Techniques in Laboratory Animals. Philadelphia, W. B. Saunders, 1968.

Greselin, E.: Detection of otitis media in the rat. Can. J. Comp. Med. Vet. Sci., *25*:274-276, 1961.

Margulis, A. R., Carlsson, E., and McAlister, W. H.: Angiography of malignant tumors in mice. Acta Radiol., *56*:179-192, 1961.

Tirman, W. S., et al.: Microradiography: its application to the study of the vascular anatomy of certain organs of the rabbit. Radiology, *57*:70-80, 1951.

Euthanasia

Scott, W. N., and Ray, P. M.: Euthanasia. *In* Handbook on the Care and Management of Laboratory Animals. Edited by UFAW. Baltimore, Williams & Wilkins, 1972.

Smith, D. C.: Methods of euthanasia and disposal of laboratory animals. *In* Methods of Animal Experimentation. Vol. 1. Edited by W. I. Gay. New York, Academic Press, 1965.

Chapter 4

Clinical Syndromes and Differential Diagnoses

This chapter contains the common signs and syndromes encountered in rabbits, guinea pigs, hamsters, gerbils, mice, and rats. These syndromes include cutaneous, gastrointestinal, respiratory, reproductive, neuromuscular, and miscellaneous conditions. The content of the chapter, divided by species, provides several differential diagnoses for each syndrome. These diagnostic alternatives, based on literature references and personal observations and communications, are not exhaustive; additional information might be extrapolated from other species or gleaned from the bibliographies.

Detailed information concerning specific diagnoses may be found through the Index; in the numbered, species-categorized bibliographies following this chapter (selected references are indicated by superscript numerals); or in the general reference works listed throughout Chapter 2.

Preceding the section on syndromes is an outline for a clinical history protocol. Information derived from this protocol will facilitate definition of the syndrome.

History Protocol

The discussion of Factors Predisposing to Disease (Chapter 1) and the topics in this listing provide a basis for establishing an information base before proceeding toward a diagnosis.

Animal Involved

Species
Sex
Age
Source
Breeding status

Environment of the Animal

Cage type and location
Feeder and waterer type
Bedding type
Sanitation level
Waste disposal
Other species or vermin in the colony
Disturbances in the colony
Recent changes or new animals
Light cycles
Ventilation
Humidity
Experimental procedure

Diet of the Animal

Source and composition
Storage facilities
Milling date
Dietary supplementation
Recent changes in diet

Health History

Previous diseases
Treatments
Breeding record
Quarantine procedure

Complaint

Onset and progression
Number exposed
Number affected
Ages and sex affected
Clinical signs

Rabbit Syndromes

Common clinical syndromes in the domestic rabbit include otitis externa, diarrhea, "slobbers," malocclusion, torticollis, nasal discharge (snuffles), conjunctivitis, infertility, posterior paralysis, weight loss, and sudden death.

CUTANEOUS SYNDROMES

- Dermatitis
  - Ectoparasitism
    - Acariasis
      - *Psoroptes cuniculi* (ear mite)
      - *Cheyletiella parasitivorax* (fur mite)
      - *Sarcoptes* and *Notoedres* (mange mites)
    - Pediculosis
      - *Haemodipsus ventricosis*
  - Ulcerative dermatitis
    - Pododermatitis
    - Bite wounds
    - Moist dermatitis
  - Bacterial dermatitis
    - *Pasteurella multocida* conjunctivitis
    - *Treponema cuniculi*
    - *Staphylococcus aureus*
    - *Pseudomonas aeruginosa*
    - *Fusobacterium necrophorum*[6]
    - *Corynebacterium* spp
  - Dermatophytosis
    - *Trichophyton mentagrophytes*
  - Myxomatosis
  - Rabbit pox[9,15]
- Cutaneous or Subcutaneous Swelling
  - Abscess
    - *Pasteurella multocida*
    - *Staphylococcus aureus*[37]
  - Mammary gland
    - Lactation
    - Mastitis
    - Mammary neoplasia
  - Parasitism
    - Cuterebriasis[5,20]
    - *Multiceps* (*Coenurus*) *serialis* cysts
  - Neoplasia
    - Shope fibroma
    - Shope papilloma
    - Lymphosarcoma
    - Mammary adenocarcinoma
    - Epidermal carcinoma
  - Hematoma
- Alopecia
  - Hair pulling
    - Nest building
    - Hair chewing
    - Human caused
  - Rubbing or scratching
    - On cage or crock
    - Pruritic response
  - Genetic hairlessness
    - Hypotrichosis
  - Dermatophytosis
  - Seasonal molt

GASTROINTESTINAL SYNDROMES

- Diarrhea
  - Nonspecific enteropathy[7,8,13,22,24,31,32,34–36,45]
  - Intestinal coccidiosis
  - Mucoid enteropathy
  - Bacterial enteritis
    - Salmonellosis
    - *Pseudomonas aeruginosa*[3,21]
    - Colibacillosis[39]

Ptyalism or Wet Dewlap
- Malocclusion[27,44,47]
- Abrasion on feeder
- Abdominal pain

Pendulous Abdomen
- Nonspecific enteropathy
- Hepatic coccidiosis
- Mammary gland
- Obesity

Polydipsia
- Lactation
- Heat stress
- Febrile disease
- Enteropathy
- Leaking water bottle
- Diabetes mellitus or insipidus

Constipation
- Gastric hairball[42]
- Enteropathy
- Anorexia
- Hypertrophic pyloric stenosis[43]

Anorexia
- Malocclusion[27,44,47]
- Illness or pain
- Insufficient water
- Heat stress
- Unpalatable feed

RESPIRATORY SYNDROMES

Nasal Discharge
- Rhinitis
  - *Pasteurella multocida*
  - *Bordetella bronchiseptica*
- Bronchopneumonia
- Heat stress
- Allergy
- Myxomatosis
- Rabbit pox[9,15]

Dyspnea
- Heat stress
- Pneumonia
  - *Pasteurella multocida*
  - *Staphylococcus aureus*
  - *Bordetella bronchiseptica*
  - *Pseudomonas aeruginosa*
- Pregnancy toxemia
- Pyothorax

Conjunctivitis
- *Pasteurella multocida*
- Dust or trauma
- Myxomatosis
- Rabbit pox

REPRODUCTIVE SYNDROMES

Infertility
- Immaturity or senescence
- Environmental stressors
  - Noise
  - Heat stress[12,26]
  - Crowding
  - Autumnal phenomenon
- Incompatible pair
- Bacterial infection
  - Endometritis
  - Pyometra
  - Orchitis
  - Venereal spirochetosis
- Endometriosis or cystic endometrial hyperplasia
- Uterine adenocarcinoma
- Nutritional deficiency
- Estrogens in feed[46]
- Pseudopregnancy[18]

Vaginal Discharge
- Normal urine
- Uterine adenocarcinoma
- Pyometra
- Abortion

Prenatal Mortality
- Nitrates in feed or water[19]
- Pregnancy toxemia
- Malnutrition[23,29]
- Congenital abnormality[23]
- Environmental noise
- Uterine adenocarcinoma
- Infectious disease
  - Pasteurellosis
  - Listeriosis[14,17,41]
  - Salmonellosis
  - Chlamydial infection
- Uterine crowding before 17 days[1,2]
- Handling trauma at 17 to 23 days[2]
- Colony crowding

Litter Desertion or Cannibalism
- Maternal inexperience

Environmental disturbances
Agalactia
Gastric hairball[42]
Malocclusion[27,44,47]
Anorexia
Mastitis
Reduced feed or water intake
Insufficient nesting material
Deformed young
Thirst
Malnutrition
Split nest

EXCITABLE TISSUES AND SPECIAL SENSES

Torticollis
Otitis interna
Encephalitis
Pasteurella multocida
Other bacteria[14,17,41]
Ascaris columnaris[10,25]
Encephalitozoonosis
Hereditary scoliosis
Incoordination or Convulsions
Traumatic injury
Otitis interna
Encephalitis[10,25]
Poisoning
Insecticides
Fertilizers
Pregnancy toxemia
Agonal phenomenon
Congenital abnormality
Magnesium deficiency
Muscular Weakness, Paresis, or Paralysis
Vertebral luxation or fracture[30,40]
Congenital abnormality
Splay leg complex[4]
Ataxia
Malnutrition[29]
Sarcocystis spp

MISCELLANEOUS SYNDROMES

Weight Loss
Malocclusion[27,44,47]
Incisors
Cheek teeth
Gastric hairball[42]
Nutritional deficiency
Chronic disease
Pasteurellosis
Coccidiosis
Neoplasia
Yersiniosis[11]
Salmonellosis
Listeriosis[14,17,41]
Pain
Ectoparasitism
Arteriosclerosis[26,33,38]
Sudden Death
Chilling or heat stress
Septicemia or toxemia
Salmonellosis
Pasteurellosis
Tularemia
Enterotoxemia
Bacillus piliformis
Pregnancy toxemia
Starvation or dehydration
Chronic disease
Litter abandoned
Congenital abnormality
Trauma
Myxomatosis
Lymphosarcoma
Rabbit pox[9,15]
Anemia
Haemodipsus ventricosis
Lymphosarcoma
Genetic abnormality

Guinea Pig Syndromes

Common clinical syndromes in the guinea pig include bite wounds, cervical abscesses, pregnancy associated alopecia, malocclusion, anorexia, pneumonia, abortion, torticollis, scorbutus, weight loss, and sudden death.

CUTANEOUS SYNDROMES

Dermatitis
Bite wounds
Dermatophytosis
Trichophyton mentagrophytes

Bacterial dermatitis
Ectoparasitism
Acariasis
Chirodiscoides caviae
Pediculosis
Gliricola porcelli
Gyropus ovalis
Cutaneous or Subcutaneous Swelling
Abscess
Streptococcus zooepidemicus
Staphylococcus aureus[11]
Streptobacillus moniliformis[1]
Yersinia pseudotuberculosis[2]
Mastitis[5]
Arthritis[4]
Scorbutus
Alopecia
Intensively bred females
Barbering
Wire abrasion

GASTROINTESTINAL SYNDROMES

Diarrhea
Nonspecific enteropathy
Acute cecitis
Antibiotic "toxicity"
Salmonellosis
Coccidiosis
Nematodiasis
Paraspidodera uncinata[9]
Ptyalism
Malocclusion
Hypovitaminosis C
Adrenocortical insufficiency

RESPIRATORY SYNDROMES

Nasal Discharge or Dyspnea
Pneumonia
Bordetella bronchiseptica
Diplococcus pneumoniae
Klebsiella pneumoniae
Streptococcus zooepidemicus
Pasteurella multocida
Pseudomonas aeruginosa
Staphylococcus aureus
Rhinitis
Heat stress
Pregnancy toxemia
Conjunctivitis
Chlamydial neonatal conjunctivitis
Streptococcus pneumoniae
Foreign body or trauma
Demodicidosis

REPRODUCTIVE SYNDROMES

Infertility
Nutritional deficiency
Environmental stress
Bedding adhering to genitalia[8]
Immaturity or senescence
Estrogens in feed[12]
Seasonal phenomenon
Prenatal Mortality[3]
Nutritional deficiency
Bacterial diseases[6]
Bordetella bronchiseptica
Salmonellosis
Streptococcus spp
Pregnancy toxemia
Environmental stress
Litter Desertion or Cannibalism
Mastitis[5]
Maternal inexperience
Environmental disturbance

EXCITABLE TISSUES AND SPECIAL SENSES

Torticollis
Otitis interna
Respiratory pathogens
Bacterial encephalitis
Incoordination or Convulsions
Pregnancy toxemia
Traumatic injury
Agonal phenomenon
Reluctance to Move or Paralysis
Hypovitaminosis C
Luxation or fracture
Chronic disease
Vitamin E deficiency
Osteoarthritis[4]

MISCELLANEOUS SYNDROMES

Weight Loss
- Malocclusion
- Nutritional deficiency
- Anorexia
- Metastatic calcification[7,10]
- Chronic disease
- Ectoparasitism
- Yersiniosis[2]

Sudden Death
- Chilling or overheating
- Septicemia or toxemia
  - Salmonellosis
  - *Streptococcus* spp
- Acute enteritis
- Pregnancy toxemia
- Starvation or dehydration
- Lymphosarcoma
- Trauma
- Antibiotic "toxicity"
- Dystocia
- Urinary calculi

Hamster Syndromes

Spontaneous diseases in the golden or Syrian hamster are, with some notable exceptions, uncommon; therefore, the differential diagnoses are limited and often stated in nonspecific terms. The more common clinical syndromes in the golden hamster include cannibalism, diarrhea (wet tail), skin wounds, dermatitis, polyuria, and sudden death.

CUTANEOUS SYNDROMES

Dermatitis
- Ectoparasitism
  - *Demodex aurati*
  - *Demodex criceti*
- Bite wounds
- Bacterial dermatitis

Cutaneous or Subcutaneous Swelling
- Cheek pouches
- Abscess
  - *Pasteurella pneumotropica*
  - *Staphylococcus aureus*
- Arthritis
- Pregnancy
- Testicles
- Mastitis[1]
- Neoplasia
- Cuterebriasis

Alopecia
- Genetic hairlessness
- Demodectic mange
- Adrenocortical neoplasia

GASTROINTESTINAL SYNDROMES

Diarrhea or Rectal Prolapse
- Proliferative ileitis
- Tyzzer's disease
- Cestodiasis
  - *Hymenolepis nana*
  - *Hymenolepis diminuta*
- Antibiotic "toxicity"[5]
- Protozoal infection
- Salmonellosis

Polydipsia and polyuria
- Cystic renal disease
- Renal amyloidosis

RESPIRATORY SYNDROMES

Nasal Discharge or Dyspnea
- Rhinitis
- Pneumonia
- Heat stress
- Sendai virus infection

Conjunctivitis
- Bacterial conjunctivitis
- Foreign body or trauma
- Lymphocytic choriomeningitis

REPRODUCTIVE SYNDROMES

Infertility
- Immaturity or senescence
- Nutritional deficiency
- Environmental stress
- Incompatible pair
- Seasonal phenomenon

Prenatal Mortality[4]
- Nutritional deficiency
- Infectious disease

- Environmental stress
- Large litter

Litter Desertion or Cannibalism[2]

- Environmental disturbance
- Maternal inexperience
- Agalactia
- Mastitis[1]
- Reduced food or water intake
- Young on wire caging

EXCITABLE TISSUES AND SPECIAL SENSES

Torticollis

- Otitis interna
- Encephalitis

Incoordination or Convulsions

- Encephalitis
    - Bacterial infection
    - Lymphocytic choriomeningitis
- Trauma

MISCELLANEOUS SYNDROMES

Weight Loss

- Malocclusion
- Chronic disease
    - Salmonellosis
    - Amyloidosis
    - Nephrosis
    - Hepatic cirrhosis
    - Neoplasia
- Nutritional Deficiency
- Gastric hairball[3]
- Parasitism
    - *Hymenolepis* spp

Sudden Death

- Starvation or dehydration
- Septicemia or toxemia
    - Proliferative ileitis
    - *Streptococcus* spp
    - Salmonellosis
    - Tularemia
- Chilling or overheating
- Chronic disease
- Litter abandoned
- Amyloidosis
- Congenital abnormality
- Antibiotic "toxicity"
- Poisoning
- Sendai virus infection
- *Bacillus piliformis*

Gerbil Syndromes

As with hamsters, spontaneous diseases are uncommon in the Mongolian gerbil. The differential diagnoses listed below, therefore, include many conditions of rare occurrence. Common clinical syndromes in the gerbil include rough hair coat, epileptiform seizures, "sore nose," mate rejection, diarrhea, and sudden death.

CUTANEOUS SYNDROMES

Dermatitis

- Trauma
    - Sore nose
- Bite wounds
- Bacterial dermatitis[6]
- Ectoparasitism
    - Demodectic mange
- Dermatophytosis

Cutaneous or Subcutaneous Swelling

- Neoplasia
- Abscess

Alopecia

- Neonatal runt syndrome
- Hair chewing with crowding

Rough Hair Coat

- Humidity over 50%
- Improper or insufficient bedding
- Leaky water bottle
- Chronic disease
    - *Bacillus piliformis*
    - Neoplasia

GASTROINTESTINAL SYNDROMES

Diarrhea

- *Bacillus piliformis*
- Nonspecific enteropathy
- Salmonellosis
- *Hymenolepis nana*

RESPIRATORY SYNDROMES

Nasal Discharge or Dyspnea
- Rhinitis
- Pneumonia
- Heat stress

Conjunctivitis
- Problem in older gerbils
- Bacterial conjunctivitis
- Foreign body or trauma

REPRODUCTIVE SYNDROMES

Infertility
- Incompatible pair
- Immaturity or senescence
- Overcrowding
- Nutritional deficiency
- Environmental disturbance
- Cystic ovaries[4]
- Neoplasia

Prenatal Mortality
- Nutritional deficiency
- Environmental stress

Litter Desertion or Cannibalism
- Small litter
- Exposed nest
- Lack of nesting material
- Agalactia
- Environmental disturbance
- Overcrowding

EXCITABLE TISSUES AND SPECIAL SENSES

Torticollis
- Otitis interna
- Encephalitis

Incoordination or Convulsions
- Epileptiform seizures[1,3,7]
- Encephalitis
- Poisoning
- Trauma

MISCELLANEOUS SYNDROMES

Weight Loss
- Nutritional deficiency
- Malocclusion[2]
- Neoplasia[5]
- *Bacillus piliformis*

Sudden Death
- Starvation or dehydration
- Septicemia or toxemia
- Chilling or overheating
- Neoplasia[5]
- Litter abandoned
- Acute gastritis, overeating

Mouse Syndromes

The determination of the incidence of disease in the mouse is complicated by the variations in the incidence of disease among the several strains of mice. Common syndromes in mice include respiratory disease, ectoparasitism, neoplasia, enteritis, weight loss, and sudden death.

CUTANEOUS SYNDROMES

Dermatitis
- Dermatophytosis
- Bite wounds
- Ulcerative dermatitis
  - Ectromelia
  - Self-mutilation
    - Ectoparasitism
    - Otitis media[7]
  - Bacterial dermatitis
  - Idiopathic in certain inbred strains[13]
- Ectoparasitism
  - Acariasis
    - *Myobia musculi*
    - *Radfordia affinis*
    - *Myocoptes musculinus*
    - *Liponyssus bacoti*
  - Pediculosis
    - *Polyplax serrata*
- Nematodiasis
  - *Syphacia obvelata* (pinworm)
  - Bedding abrasion

Cutaneous or Subcutaneous Swelling
- Follicular mite
  - *Psorergates simplex*
- Estrogen induced scrotal hernia
- Neoplasia
- Abscess
  - *Pasteurella pneumotropica*
  - *Corynebacterium kutscheri*
- Pregnancy

Alopecia
- Abrasion[9]
- Bite wounds
- Barbering
- Idiopathic alopecia
- Endocrine imbalance
- Reovirus infection[3]

Appendage Inflammation or Amputation
- Ectromelia
- *Streptobacillus moniliformis*[5]
- Fighting
- Arthritis
  - *Mycoplasma arthritidis*
  - *Corynebacterium kutscheri*

Rough Hair Coat
- Senility
- Debilitation
- Dirty cage
- Diarrhea
- Leaky water bottle

GASTROINTESTINAL SYNDROMES

Diarrhea
- Viral diarrhea in unweaned mice
  - Epizootic diarrhea of infant mice (EDIM)
  - Reovirus infection[3]
  - Mouse hepatitis virus (MHV)
- Salmonellosis
- *Citrobacter freundii*[2]
- *Bacillus piliformis*
- Dietary factors
- Hyperplastic colitis
- Hexamitiasis[11]

Pendulous Abdomen
- Pregnancy
- Hexamitiasis
- Autoimmune renal disease
- Ascites
- Cardiomyopathy
- Salmonellosis
- Neoplasia
- *Cysticercus fasciolarus*

Prolapsed Rectum
- Hyperplastic colitis
- Pinworms
- AKR syndrome
- Estrogenic stimulation

RESPIRATORY SYNDROMES

Nasal Discharge
- *Pasteurella pneumotropica*
- Dust

Pneumonia
- *Mycoplasma* spp
- *Pasteurella pneumotropica*
- Sendai virus infection
- *Klebsiella pneumoniae*
- *Corynebacterium kutscheri*
- *Bordetella bronchiseptica*
- Pneumonia virus of mice (PVM)
- K virus infection

Conjunctivitis
- Dacryoadenitis
- *Corynebacterium kutscheri*
- *Mycoplasma* spp
- *Bordetella bronchiseptica*

REPRODUCTIVE SYNDROMES

Infertility
- Improper light cycle
- Immaturity or senescence
- Overcrowding
- Inadequate bedding
- Cystic ovaries
- Inbreeding
- Chemicals[10,14]

Vaginal Discharge
- Postcopulation plug
- Estrus
- Urogenital infection

Litter Desertion or Cannibalism
- Environmental disturbances
- Lack of nesting material
- Dead or deformed young
- Agalactia
- Small litter
- Inexperienced dam

EXCITABLE TISSUES AND SPECIAL SENSES

Torticollis
- Otitis interna

Incoordination or Convulsions
- Otitis interna
- *Mycoplasma neurolyticum*
- Mouse hepatitis virus (MHV)
- Trauma
- Audiogenic seizures in DBA mice[8]
- Neoplasia

Paresis or Paralysis
- Polioencephalomyelitis
- Trauma

MISCELLANEOUS SYNDROMES

Weight Loss
- Ectoparasitism
- *Hymenolepis* spp
- Neoplasia
- Malocclusion
- Malnutrition
- Salmonellosis
- Epizootic diarrhea of infant mice
- Pinworms
- *Bacillus piliformis*
- Subclinical viral infections
- Autoimmune disease

Sudden Death
- Starvation or dehydration
- Heat stress
- Malnutrition
- Trauma
- Poisoning
  - Organophosphate[14]
  - Chloroform[4]
  - Streptomycin[6]
  - Procaine
- Hexamitiasis[11]
- Giardiasis
- Bacterial septicemia
  - Salmonellosis
  - Pasteurellosis
  - *Pseudomonas aeruginosa*[1,15]
  - *Streptococcus* spp
- Viral infection
  - Ectromelia
  - Sendai virus infection
  - Adenovirus infection
  - Mouse hepatitis virus (MHV)

Ptyalism
- Heat stress
- Malocclusion

Anemia
- Pediculosis
- Leukemia
- Autoimmune disease
- *Eperythrozoon coccoides*[12]

Rat Syndromes

The laboratory rat has few common spontaneous diseases, but those diseases which do occur are common and debilitating over a chronic course. Rat diseases frequently seen are "ringtail," mammary neoplasia, chronic respiratory disease, nephrosis, and a dacryoadenitis with porphyria.

CUTANEOUS SYNDROMES

Dermatitis
- Dermatophytosis
- Ulcerative dermatitis[1]
- Ectoparasitism
  - *Polyplax spinulosa*
  - *Liponyssive bacoti*
- Bite wounds
- Conjunctivitis

Cutaneous or Subcutaneous Swelling
- Abscess
- Mammary neoplasia
- Viral sialodacryoadenitis[9]
- Arthritis[12]

Alopecia
- Dermatophytosis
- Trauma or abrasion
- Excessive grooming[2]

Limb or Tail Necrosis
- Ringtail[3,11,14]
- Trauma or thrombosis

Rough Hair Coat
- Common in older rats
- Malnutrition
- Febrile disease

Conjunctivitis
- Viral sialodacryoadenitis[9]
- Chromodacryoadenitis
- Mycoplasmosis
- Trauma

GASTROINTESTINAL SYNDROMES

Diarrhea
- Nonspecific enteropathy
- Neurogenic

Pendulous Abdomen
- Pregnancy
- Megaloileitis
- Obesity

RESPIRATORY SYNDROMES

Nasal Discharge or Dyspnea
- Sialodacryoadenitis virus
- *Mycoplasma pulmonis*
- *Pasteurella pneumotropica*
- *Streptococcus pneumoniae*
- *Bordetella bronchiseptica*
- *Corynebacterium kutscheri*
- *Haemophilus* spp[7]

REPRODUCTIVE SYNDROMES

Infertility
- Immaturity or senescence
- Vitamin E deficiency
- Organophosphate poisoning[13]
- Ambient temperature and humidity[11]
- Improper light intensity or cycle
- Ectoparasitism

Litter Desertion or Cannibalism[10]
- Excessive noise
- Deformed or dead young
- Overcrowding
- Abrasion of pup's skin
- Male present at parturition
- Environmental stress
- Small litter
- Dirty cage
- Lack of nesting material

EXCITABLE TISSUES AND SPECIAL SENSES

Torticollis
- Otitis interna
- Pituitary adenoma
- Encephalitis

Incoordination or Convulsions
- Trauma
- Encephalitis
- Neoplasia
- Kilham rat virus
- Otitis interna

Paresis or Paralysis
- Brain lesion
- Trauma
- Malnutrition
- Arthritis[12]

MISCELLANEOUS SYNDROMES

Weight Loss
- Overcrowding
- Malocclusion
- Ectoparasitism
  - *Liponyssus bacoti*
  - *Polyplax spinulosa*
- *Hymenolepis* spp
- Malnutrition
- Nephrosis

Sudden Death
- Overheating or chilling
- Trauma
- Septicemia
  - *Streptococcus pneumoniae*
  - *Pasteurella pneumotropica*
  - *Corynebacterium kutscheri*
- Malnutrition
- Salmonellosis
- Megaloileitis[6,8]
- Hemorrhagic enteritis
- Stress

Ptyalism
- Suffocation
- Heat stress
- Malocclusion

Anemia
- Ectoparasitism
  - *Liponyssus bacoti*
  - *Polyplax spinulosa*
- Hemobartonellosis[4]

BIBLIOGRAPHY

ILAR Committee on Laboratory Animal Diseases: A Guide to Infectious Diseases of Guinea Pigs, Gerbils, Hamsters, and Rabbits. Washington, DC, National Academy of Sciences, 1974.

ILAR Committee on Laboratory Animal Diseases: A Guide to Infectious Diseases of Mice and Rats.

Washington, DC, National Academy of Sciences, 1971.

Schuchman, S. M.: Individual care and treatment of mice, rats, guinea pigs, hamsters, and gerbils. *In* Current Veterinary Therapy V. Edited by R. W. Kirk. Philadelphia, W. B. Saunders Co., 1974.

Soave, O. A.: Diagnosis and control of common diseases of hamsters, rabbits, and monkeys. J. Am. Vet. Med. Assoc., *142*:285-290, 1963.

Trum, B. F., and Routledge, J. K.: Common disease problems in laboratory animals. J. Am. Vet. Med. Assoc., *151*:1886-1896, 1967.

Rabbit Syndromes

1. Adams, C. E.: Maintenance of pregnancy relative to the presence of few embryos in the rabbit. J. Endocrinol., *48*:243-249, 1970.
2. Adams, C. E.: Studies on prenatal mortality in the rabbit, *Oryctolagus cuniculus*: The amount and distribution of loss before and after implantation. J. Endocrinol., *19*:325-344, 1960.
3. Alpen, G. R., and Maerz, K.: The incidence of a pathogenic strain of *Pseudomonas* in a rabbit colony. J. Inst. Anim. Tech., *20*:72-74, 1969.
4. Arendar, G. M., and Milch, R. A.: Splay leg—a recessively inherited form of femoral neck anterversion, femoral shaft torsion and subluxation of the hip in the laboratory lop rabbit. Clin. Ortho., *44*:221-229, 1966.
5. Beamer, R. H., and Penner, L. R.: Observations on the life history of a rabbit cuterebrid, the larvae of which may penetrate the human skin. J. Parasitol., *28*:25, 1942.
6. Beattie, J. M., Gates, A. G., and Donaldson, M. A.: An epidemic disease in rabbits resembling that produced by *B. necrosis* (Schmorl) but caused by an aerobic bacillus. J. Pathol. Bacteriol., *18*:34-36, 1913.
7. Bryner, J. H., et al.: Infectivity of three *Vibrio fetus* biotypes for gallbladder and intestines of cattle, sheep, rabbits, guinea pigs, and mice. Am. J. Vet. Res., *32*:465-470, 1971.
8. Casady, R. B., Hagen, K. W., and Sittman, K.: Effects of high level antibiotic supplementation in the ration on growth and enteritis in young domestic rabbits. J. Anim. Sci., *23*:477-480, 1964.
9. Christensen, L. R., Bond, E., and Matanic, B.: Pockless rabbit pox. Lab. Anim. Care, *17*:281-296, 1967.
10. Dade, A. W., et al.: An epizootic of cerebral nematodiasis in rabbits due to *Ascaris columnaris*. Lab. Anim. Sci., *25*:65-69, 1975.
11. Daniels, J. J. H. M.: Enteral infections with *Pasteurella pseudotuberculosis*. Br. Med. J., *2*:997, 1961.
12. Dogget, V. C.: Periodicity in the fecundity of male rabbits. Am. J. Physiol., *187*:445-450, 1956.
13. Duncan, C. L., and Strong, D. H.: Experimental production of diarrhea in rabbits with *Clostridium perfringens*. Can. J. Microbiol., *15*:765-816, 1969.
14. Gray, M. L., Singh, C., and Thorp, F.: Abortions, stillbirths, early death of young in rabbits by *Listeria monocytogenes*. II. Oral exposure. Proc. Soc. Exp. Biol. Med., *89*:169-175, 1955.
15. Green, H. S. N.: Rabbit pox. I. Clinical manifestations and course of the disease. J. Exp. Med., *60*:427-440, 1934.
16. Haust, M. D., and Greer, J. C.: Mechanism of calcification in spontaneous aortic arteriosclerotic lesions of the rabbit. Am. J. Pathol., *60*: 329-346, 1970.
17. Holmes, R. G.: Listeriosis in rabbits. Vet. Rec., *73*:791, 1961.
18. Kaufmann, A. F., et al.: Pseudopregnancy in the New Zealand white rabbit: necropsy findings. Lab. Anim. Sci., *21*:865-869, 1971.
19. Kruckenberg, S. M.: Nitrate induced abortions in rabbits: observations of field and laboratory cases. Abst. #23, 25th Annual Session, AALAS, Cincinnati, 1974.
20. Lopushinsky, T.: Myiasis of nesting cottontail rabbit. J. Wildl. Dis., *6*:98-100, 1970.
21. McDonald, R. A., and Pinheiro, A. F.: Water chlorination controls *Pseudomonas aeruginosa* in a rabbitry. J. Am. Vet. Med. Assoc., *151*:863-864, 1967.
22. Mack, R.: Disorders of the digestive tract of domesticated rabbits. Vet. Bull., *32*:1-8, 1965.
23. Millen, J. W., and Dickson, A. D.: The effect of vitamin A upon the cerebrospinal fluid pressures of young rabbits suffering from hydrocephalus due to maternal hypovitaminosis. Br. J. Nutr., *11*:440-446, 1957.
24. Moon, H. W., et al.: Intraepithelial *Vibrio* associated with acute typhlitis of young rabbits. Vet. Pathol., *11*:313-329, 1974.
25. Nettles, V. F., et al.: An epizootic of cerebrospinal nematodiasis in cottontail rabbits. J. Am. Vet. Med. Assoc., *167*:600-604, 1975.
26. Oloufa, M. M., Bogart, R., and McKenzie, F. F.: Effect of environmental temperatures and the thyroid gland on fertility in the male rabbit. Fertil. Steril., *2*:223-228, 1951.
27. Pollock, S.: Slobbers in the rabbit. J. Am. Vet. Med. Assoc., *119*:443-444, 1951.
28. Renquist, D., and Soave, O.: Staphylococcal pneumonia in a laboratory rabbit: an epidemiologic follow-up study. J. Am. Vet. Med. Assoc., *155*:1221-1223, 1969.
29. Ringler, D. H., and Abrams, G. D.: Nutritional muscular dystrophy and neonatal mortality in a rabbit breeding colony. J. Am. Vet. Med. Assoc., *157*:1928-1934, 1970.
30. Roe, F. J. C., and Stiff, A. L.: Fracture dislocation of lumbar spine occurring spontaneously in rabbits. J. Anim. Tech. Assoc., *12*:92-94, 1962.
31. Rollins, W. C., and Casady, R. B.: An analysis of preweaning deaths in rabbits with special emphasis on enteritis and pneumonia. Anim. Prod., *9*:87-92, 1967.
32. Savage, M. L., et al.: An epizootic of diarrhea in a rabbit colony. Pathology and bacteriology. Can. J. Comp. Med., *37*:313-319, 1973.
33. Schenk, E. A., Gaman, E., and Feigenbaum,

A. S.: Spontaneous aortic lesions in rabbits. I. Morphologic characteristics. Circ. Res., *19*:80-88, 1966.
34. Smith, H. W.: Observations on the flora of the alimentary tract of animals and factors affecting its composition. J. Pathol. Bacteriol., *89*:95, 1965.
35. Smith, H. W.: The antimicrobial activity of the stomach contents of suckling rabbits. J. Pathol. Bacteriol., *91*:1-9, 1966.
36. Smith, H. W.: The development of the flora of the alimentary tract in young animals. J. Pathol. Bacteriol., *90*:495-513, 1965.
37. Snyder, S. B., et al.: Disseminated staphylococcal disease in laboratory rabbits (*Oryctolagus cuniculus*). Lab. Anim. Sci., *26*:1, 1976.
38. Stevenson, R. G., Palmer, N. C., and Finley, G. G.: Hypervitaminosis D in rabbits. Can. Vet. J., *17*:54-57, 1976.
39. Taylor, J., Williams, M. P., and Payne, J.: Relation of rabbit gut reaction to enteropathogenic *Escherichia coli*. Br. J. Exp. Pathol., *42*:43-52, 1961.
40. Templeton, G. S.: Treatment for paralyzed hindquarters. Am. Rabbit J., *16*:155, 1946.
41. Vetesi, F., and Kemeres, F.: Studies on listeriosis in pregnant rabbits. Acta Vet. Acad. Sci. Hung., *17*:27-38, 1967.
42. Wagner, J. L., Hackel, D. B., and Samsell, A. G.: Spontaneous deaths in rabbits resulting from gastric trichobezoars. Lab. Anim. Sci., *24*:826-830, 1974.
43. Weisbroth, S. H., and Scher, S.: Naturally occurring hypertrophic pyloric stenosis in the domestic rabbit. Lab. Anim. Sci., *25*:355-360, 1975.
44. Weisbroth, S. H., and Ehrman, L.: Malocclusion of the rabbit. J. Hered., *58*:245-246, 1967.
45. Whitney, J. C.: Treatment of enteric disease in the rabbit. Vet. Rec., *95*:533, 1974.
46. Wright, P. A.: Infertility in rabbits by feeding ladino clover. Proc. Soc. Exp. Biol. Med., *105*:428-430, 1960.
47. Zeman, W. V., and Fielder, F. G.: Dental malocclusion and overgrowth in rabbits. J. Am. Vet. Med. Assoc., *155*:1115-1119, 1969.

Guinea Pig Syndromes

1. Alfred, P., et al.: The isolation of *Streptobacillus moniliformis* from the cervical abscesses of guinea pigs. Lab. Anim., *8*:275-277, 1974.
2. Bishop, L. M.: Study of an outbreak of pseudotuberculosis in guinea pigs (cavies) due to *B. pseudotuberculosis rodentium*. Cornell Vet., *22*:1-9, 1932.
3. Edwards, J. J.: Prenatal loss of fetuses and abortion in guinea pigs. Nature, *210*:223-224, 1960.
4. Gupta, B. N., Conner, G. H., and Meyer, D. B.: Osteoarthritis in guinea pigs. Lab. Anim. Sci., *22*:362-368, 1972.
5. Gupta, B. M., Langham, R. F., and Conner, G. H.: Mastitis in guinea pigs. Am. J. Vet. Res., *31*:1703-1707, 1970.
6. Juhr, N. C., and Obi, S.: Uterine infection in guinea pigs. Z. Versuchstierkd., *12*:383-387, 1970.
7. Maynard, L. A., et al.: Dietary mineral interrelationships as a cause of soft tissue calcification in guinea pigs. J. Nutr., *64*:85-97, 1958.
8. Plank, J. S., and Irwin, R.: Infertility of guinea pigs on sawdust bedding. Lab. Anim. Care, *16*:9-11, 1966.
9. Porter, D. A., and Otto, G. F.: The guinea pig nematode, *Paraspidodera uncinata*. J. Parasitol., *20*:323, 1934.
10. Sparschu, G. L., and Christie, R. J.: Metastatic calcification in a guinea pig colony: a pathologic survey. Lab. Anim. Care, *18*:520-526, 1968.
11. Taylor, J. L., et al.: Chronic pododermatitis in guinea pigs, a case report. Lab. Anim. Sci., *21*:944-945, 1971.
12. Wright, J. E., and Seibold, J. R.: Estrogen contamination of pelleted feed for laboratory animals: effect on guinea pig reproduction. J. Am. Vet. Med. Assoc., *132*:258-261, 1958.

Hamster Syndromes

1. Frisk, C. S., Wagner, J. E., and Owens, D. R.: Streptococcal mastitis in golden hamsters. Lab. Anim. Sci., *26*:97-98, 1976.
2. Haley, J.: Cannibalism in hamsters. Am. Small Stock Farmer, *35*:10-11, 1965.
3. Nelson, W. B.: Fatal hairball in a long haired hamster. Vet. Med. Small Anim. Clin., *70*:1193, 1975.
4. Purdy, R. H., and Hillemann, H. H.: Prenatal mortality in the golden hamster (*Cricetus auratus*). Anat. Rec., *106*:577-583, 1950.
5. Small, J. D.: Fatal enterocolitis in hamsters given lincomycin hydrochloride. Lab. Anim. Care, 18:411-420, 1968.

Gerbil Syndromes

1. Kaplan, H., et al.: Development of seizures in the Mongolian gerbil (*Meriones unguiculatus*). J. Comp. Physiol., *81*:267-273, 1972.
2. Loew, F. M.: A case of overgrown mandibular incisors in a Mongolian gerbil. Lab. Anim. Care, *17*:137-139, 1967.
3. Loskota, W. J., et al.: The gerbil as a model for the study of the epilepsies. Seizure patterns and ontogenesis. Epilepsia, *15*:109-119, 1974.
4. Norris, M. L., and Adams, C. E.: Incidence of cystic ovaries and reproductive performance in the Mongolian gerbil, *Meriones unguiculatus*. Lab. Anim., *6*:337-342, 1972.
5. Norris, M. L., and Adams, C. E.: Mortality from birth to weaning in the Mongolian gerbil. Lab. Anim., *6*:49-53, 1972.
6. Peckham, J. C., et al.: Staphylococcal dermatitis in Mongolian gerbils (*Meriones unguiculatus*). Lab. Amin. Sci., *24*:43-47, 1974.
7. Thiessen, D. D., Lindzey, G., Friend, H. C.: Spontaneous seizures in the Mongolian gerbil, *Meriones unguiculatus*. Psychom. Sci., *11*:227-228, 1968.

Mouse Syndromes

1. Beck, R. W.: The control of *Pseudomonas aeruginosa* in a mouse breeding colony by the use of chlorine in the drinking water. Lab. Anim. Care, *13*:41-45, 1963.
2. Brennan, P. C., et al.: *Citrobacter freundii* associated with diarrhea in laboratory mice. Lab. Anim. Care, *15*:266-275, 1965.
3. Cook, I.: Reovirus type 3 infection in laboratory mice. Aust. J. Exp. Biol. Med. Sci., *41*:651-659, 1963.
4. Deringer, M. K., Dunn, T. B., and Heston, W. E.: Results of exposure of strain C3H mice to chloroform. Proc. Soc. Exp. Biol., *83*:474-479, 1953.
5. Freundt, E. A.: Arthritis caused by *Streptobacillus monoliformis* and pleuropneumonia-like organisms in small rodents. Lab. Invest., *9*:1358-1375, 1959.
6. Galloway, J. H.: Antibiotic toxicity in white mice. Lab. Anim., Care, *18*:421-425, 1968.
7. Harkness, J. E., and Wagner, J. E.: Self mutilation in mice associated with otitis media. Lab. Anim. Sci., *25*:315-318, 1975.
8. Iturrian, W. B., and Fink, G. B.: Effect of noise in the animal house on seizure susceptibility and growth of mice. Lab. Anim. Care, *18*:557-560, 1968.
9. Litterst, C. L.: Mechanically self-induced muzzle alopecia in mice. Lab. Anim. Sci., *24*:806-809, 1974.
10. Les, E. P.: Effect of acidified chlorinated water on reproduction of C3H/HeJ and C57B1/6J mice. Lab. Anim. Care, *18*:210-213, 1968.
11. Lussier, G., and Loew, F. M.: An outbreak of hexamitiasis in laboratory mice. Can. J. Comp. Med., *34*:350-353, 1970.
12. Riley, V.: *Eperythrozoon coccoides*. Science, *146*:921-923, 1964.
13. Stowe, H. D., Wagner, J. L., and Pick, J. R.: A debilitating fatal murine dermatitis. Lab. Anim. Sci., *21*:892-897, 1971.
14. Wagner, J. E., and Johnson, D. R.: Toxicity of Dichlorvos for laboratory mice LD_{50} and effect on serum cholinesterase. Lab. Anim. Care, *20*:45-47, 1970.
15. Wensinck, F., Ven Bekku, D. W., and Renaud, H.: The prevention of *Pseudomonas aeruginosa* infection in irradiated mice and rats. Radiat. Res., *7*:491-499, 1957.

Rat Syndromes

1. Ash, G. W.: An epidemic of chronic skin ulceration in rats. Lab. Anim., *5*:115-122, 1971.
2. Beare-Rogers, J. L., and McGowan, J. E.: Alopecia in rats housed in groups. Lab. Anim., *7*:237-238, 1973.
3. Dikshit, P. K., and Sriramachari, S.: Caudal necrosis in suckling rats. Nature, *181*:63-64, 1958.
4. Ford, A. C., Jr., and Murray, T. J.: Studies on *Haemobartonella* infection in the rat. Can. J. Microbiol., *5*:345-350, 1959.
5. Foster, H. L.: Comparison of epizootic diarrhea of suckling rats and a similar condition in mice. J. Am. Vet. Med. Assoc., *133*:198-201, 1958.
6. Geil, R. G., Davis, C. L., and Thompson, S. W.: Spontaneous ileitis in rats—a report of 64 cases. Am. J. Vet. Res., *22*:932-936, 1961.
7. Harr, J. R., Tinsley, I. J., and Weswig, P. H.: *Haemophilus* isolated from a rat respiratory epizootic. J. Am. Vet. Med. Assoc., *155*:1126-1130, 1969.
8. Hottendorf, G. H., Hirth, R. S., and Peer, R. L.: Megaloileitis in rats. J. Am. Vet. Med. Assoc., *155*:1131 1135, 1969.
9. Jacoby, R. O., Bhatt, P. N., and Jonas, A. M.: Pathogenesis of sialodacryoadenitis in gnotobiotic rats. Vet. Pathol., *12*:196-209, 1975.
10. Mohan, C.: Age dependent cannibalism in a colony of albino rats. Lab. Anim., *8*:83-84, 1974.
11. Njaa, L. R., Utne, F., and Braekkan, O. R.: Effect of relative humidity on rat breeding and ringtail. Nature, *180*:290, 1957.
12. Skold, B. H.: Chronic arthritis in the laboratory rat. J. Am. Vet. Med. Assoc., *138*:204-207, 1961.
13. Timmons, E. H.: Dichlorvos effects on estrous cycle onset in the rat. Lab. Anim. Sci., *25*:45-47, 1975.
14. Totton, M.: Ringtail in new born Norway rats. A study of the effect of environmental temperature and humidity on incidence. J. Hyg., *56*:190-196, 1958.

Chapter 5

Specific Diseases

The diseases of laboratory animals described in this chapter were selected for inclusion because of their prevalence in pet animals, public health significance, or potential importance to research personnel. This listing of diseases is certainly not complete. Tuberculosis, listeriosis, yersiniosis, mycotic infections, many viral diseases, congenital abnormalities, and several nutritional conditions are not mentioned, except in the references. For the conditions that are included, hosts, etiologies, transmission, predisposing factors, clinical signs, necropsy signs, diagnosis, treatments, prevention, and public health significance are discussed. A bibliography follows each condition and the section.

Acariasis

HOSTS

Mites, as a group, may infect all common laboratory animals. Species of mites are usually host specific or limited to a narrow host range.

ETIOLOGY

Psoroptes cuniculi, the rabbit ear mite
Cheyletiella parasitivorax, the rabbit fur mite
Chirodiscoides caviae, the guinea pig fur mite
Demodex aurati, the hamster follicular mite
Demodex criceti, the hamster epidermal mite
Demodex spp, the gerbil demodectic mite
Myobia musculi, a mouse fur mite
Radfordia affinis, a mouse fur mite
Myocoptes musculinus, the myocoptic mange mite
Psorergates simplex, the mouse follicular mite
Liponyssus bacoti, the tropical rat mite

TRANSMISSION

Transmission of mites is by direct contact with infected hosts, crusts and scabs, or bedding.

PREDISPOSING FACTORS

Mite infestations are often more severe in debilitated hosts or in animals housed in unsanitary environments. Pet rodents attract wild rodents of similar species. This contact provides the opportunity for the transfer of many diseases, including mite infections, to domestic rodents.

CLINICAL SIGNS

Psoroptes cuniculi causes serum and crust accumulation in the external ear canal and, rarely, the adjacent head and neck of the rabbit. The inflammation and pruritus, particularly pronounced with secondary bacterial infections, may cause scratching and head shaking. Extension of the primary mite infection to the middle ear is rare, if it occurs at all, but a bacterial otitis externa may spread to the middle and inner ears.

Cheyletiella, Chirodiscoides, Myobia, Radfordia, and *Myocoptes* infections usually cause no overt signs; but in repeated infestations a partial alopecia, epidermal scale, and pruritus with self-traumatization may result.

Demodex infections are often inapparent and self-limiting; but in young, old, or debilitated hosts the infections may be more severe with a moderate to severe alopecia and dry epidermal scale and scabs over the rump and back.

Psorergates inhabits the hair follicles, and, as mites and debris accumulate, the follicular invaginations become visible as small (2 mm) white nodules on the skin of the head and neck. Alopecia may also occur with *Psorergates* infection.

Liponyssus bacoti differs from the other mites listed in that most of the life cycle is spent off an animal host. The protonymph and the adults intermittently suck blood from the host, and the loss of blood may cause anemia, decreased fertility, weakness, and death. Mammalian hosts may develop an allergic dermatitis to mites, and the pruritic response can become intense.

DIAGNOSIS

A skin scraping, deep if *Demodex* is suspected, with an oil- or glycerin-wetted scalpel blade will collect mites or lice. Scraped material is placed on a slide with a few drops of warmed 10% KOH, covered, allowed to sit 10 minutes, and then examined microscopically. The pelt, especially the head, ears, neck, and margins of lesions, can be examined with a hand lens or dissecting microscope for adult or immature ectoparasites. Engorged *Liponyssus* and other mites moving in the fur can be seen with the unaided eye.

If the animal is dead, the hand lens examination procedure will provide a more certain diagnosis if the pelt is cooled in the refrigerator for 30 minutes, removed for 10 minutes, and then examined with a lens. The parasites migrate off the cold pelt and toward the warmer tips of the hair. If the pelt is placed on a dark colored paper and surrounded by a frame of double gummed cellophane tape, the mites will become trapped after they leave the animal.

TREATMENT

Psoroptes cuniculi, the rabbit ear mite, is treated with one of several available oil-based insecticide preparations. One or 2 ml of the liquid are massaged into the ear canal. Treatment is repeated at 7 and 14 days to eliminate newly hatched mites. In severe cases the use of a sebumlytic compound, antiseptic soap, and topical or systemic antibiotic may be indicated.

Cheyletiella, Chirodiscoides, Myobia, Radfordia, Myocoptes, Psorergates, and *Liponyssus* can be eliminated by dusting the animals and bedding with a carbaryl or pyrethrin dusting powder. Individual animals may be dipped into 0.025% lindane solution, 2% malathion, or other ectoparasite dip. Preweanling animals should not be dipped. A mechanical agent that abrades and desiccates ectoparasites is a silica aerogel such as Dri-Die.[1] This compound is mixed with the contact bedding in the affected animal's cage.

Demodex spp on hamsters and gerbils burrow into the skin and are difficult to detect and treat. A 1 to 5.5 ronnel concentrate and propylene glycol mixture applied daily for 5 weeks is a suggested treatment for *Demodex*.

Resin strips containing dichlorvos placed on or near the cage for 24-hour periods at several weekly intervals will reduce ectoparasite populations in a colony. Entire rooms of animals can be treated by restricting air flow and using several resin vapor strips. Elimination of ectoparasites from the premises requires treatment and removal of the animals from the room, thorough mechanical scrubbing of equipment, and fumigation of the room with formaldehyde. Fomites (feed bags, carts, trash cans, clothing) moved from room to room must also be cleaned to remove ectoparasites. Rooms should be repopulated with animals proven free of ectoparasites.

PREVENTION

Mite infestations are prevented by placing clean stock into clean premises. Infected animals should be separated and treated, and the premises should be disinfected. Mites die within 3 weeks off the host. Wild rodents must be excluded from the colony.

PUBLIC HEALTH SIGNIFICANCE

Psoroptes, Chirodiscoides, Psorergates, Myobia, Radfordia, Myocoptes, and *Demodex* mites of rabbits and rodents are not known to affect man.

Cheyletiella parasitivorax can, in rare cases, cause a dermatitis in man.

Sarcoptes spp and *Notoedres* spp have a wide host range, including rabbits, rodents, and man. These mange mites are rarely encountered on research animals.

Liponyssus bacoti bites man and can serve as a vector for murine typhus, Q fever, and plague.

BIBLIOGRAPHY

Baies, A., Suteu, I., Klemm, W.: *Notoedres* scabies of the golden hamster. Z. Versuchsteirkd., 1:251-257, 1968.

Barr, A. R.: A case of "mange" of the domestic rabbit due to *Cheyletiella parasitivorax*. J. Parasitol., *41*:323, 1955.

Campbell, D. J.: Parasitic diseases of laboratory animals. Can. Med. Assoc. J., *98*:908-910, 1968.

Estes, P. C., Richter, C. B., and Franklin, J. A.: Demodectic mange in the golden hamster. Lab. Anim. Sci., *21*:825-828, 1971.

Flynn, R. J.: Parasites of Laboratory Animals. Ames, The Iowa State University Press, 1973.

Gibson, T. E.: Parasites of laboratory animals transmissible to man. Lab. Anim., *1*:17-24, 1967.

Griffiths, H. J.: Some common parasites of small laboratory animals. Lab. Anim., 5:123-125, 1971.

Henderson, J. D.: Treatment of cutaneous acariasis in the guinea pig. J. Am. Vet. Med. Assoc., *163*:591-592, 1973.

Lund, E. E.: Ear mange in domestic rabbits. Am. Rabbit J., *21*:67-69, 1951.

Nutting, W. B.: *Demodex aurati* and *D. criceti*, ectoparasites of the golden hamster (*Mesocricetus auratus*). Parasitol., *51*:515-522, 1968.

Ronald, N. C., and Wagner, J. E.: The arthropod parasites of the genus Cavia. *In* The Biology of the Guinea Pig. Edited by J. E. Wagner and P. J. Manning. New York, Academic Press, 1976.

Schwarzbrott, S. S., Wagner, J. E., and Frisk, C. S.: Demodicosis in the mongolian gerbil (*Meriones unguiculatus*): a case report. Lab. Anim. Sci., *24*:666-668, 1974.

Scott, H. G.: Control of mites on the hamster. J. Econ. Entomol., *51*:412-413, 1958.

Specht, F.: Inflammation of the middle ear, labyrinth, and meninges in the rabbit resulting from mites in the auditory canal. Arch. Ohren Nasen Kehlkopfheik., *128*:103-114, 1931.

Tarshis, I. B.: The use of silica aerogel compounds for the control of ectoparasites. Proc. Anim. Care Panel, *12*:217-258, 1962.

Wagner, J. E.: Control of mouse parasites with resin vapor strips containing vapona. Lab. Anim. Care, *19*:804-807, 1969.

Wagner, J. E., Al-Rabiai, S., and Rings, R. W.: *Chirodiscoides caviae* infestation in guinea pigs. Lab. Anim. Sci., *22*:750-752, 1972.

---

[1]FMC Corporation, Agricultural Chemical Division, 100 Niagara St., Middleport, NY 14105.

Weisbroth, S. H., et al.: The parasitic ecology of the rodent mite *Myobia musculi*. I. Grooming factors. Lab. Amin. Sci., *24*:510-516, 1974.

Bacillus piliformis Infection (Tyzzer's Disease)

HOSTS

Among the common laboratory animal species, rabbits, hamsters, gerbils, mice, and rats have been reported with Tyzzer's disease. It is one of the most serious diseases found in gerbils. The disease is also endemic in some mouse colonies. Other species affected include muskrats, cats, horses, and a Rhesus monkey.

ETIOLOGY

Bacillus piliformis, a Gram-negative, pleomorphic, filamentous organism and intracellular parasite, is the suspected causative organism of Tyzzer's disease.

TRANSMISSION

Transmission is fecal-oral, although infectious, sporelike bodies may survive up to one year in the bedding. Old, more resistant animals may act as carriers.

PREDISPOSING FACTORS

Poor husbandry standards and environmental stressors contribute to the development of the clinical disease, which is more common in weanling mice and gerbils. Sulfaquinoxaline used to prevent coccidiosis and pneumonia in young rabbits may precipitate an epidemic of Tyzzer's disease. Corticosteroids, sulfa drugs, and radiation can also convert a latent infection to the clinical disease.

CLINICAL SIGNS

Tyzzer's disease, usually seen in weanling or stressed animals, is an acute, enzootic disease with diarrhea, dehydration, debility and death occurring often within 48 hours. Carriers and sporadic outbreaks occur. Chronically infected animals exhibit weight loss and weakness.

NECROPSY SIGNS

The most consistent lesion of Tyzzer's disease is an enlarged liver with numerous gray and yellow foci 1 to 2 mm in diameter. The intestine, particularly in the area of the colic-cecal-ileal junction, may exhibit edema and hemorrhage. In rabbits and rats pale myocardial foci have been reported.

DIAGNOSIS

Necropsy signs and Giemsa, PAS, or silver staining of the intracellular, filamentous *B. piliformis* organism in hepatocytes provide a definitive diagnosis of Tyzzer's disease. The organism has not been cultured on cell-free media. Indirect fluorescent antibody tests are also a diagnostic aid.

TREATMENT

The acute (1 to 4 day) course of the disease, the production of resistant "spores," and the intracellular location of the parasite reduce the effectiveness of treatment. Oxytetracycline in the water at 0.1 mg/ml for 30 days has suppressed an outbreak.

PREVENTION

Clean stock and good husbandry practices are the best preventive measures. Antibiotics may suppress active infections, but carriers develop. The "spores" survive freeze-thaw but are killed if heated at 56°C for 1 hour. Filter cage covers aid in reducing transmission. A formalin-killed bacterin has been prepared but is not widely used.

PUBLIC HEALTH SIGNIFICANCE

No public health significance is known, but the report of a *B. piliformis* infection in a Rhesus monkey should be noted.

BIBLIOGRAPHY

Allen, A. M., et al.: Tyzzer's disease syndrome in laboratory rabbits. Am. J. Pathol., *46*:859-882, 1965.

Carter, G. R., Whitenack, D. L., and Julius, L. A.: Natural Tyzzer's disease in Mongolian gerbils (*Meriones unguiculatus*). Lab. Anim. Care, *19*:648-651, 1969.

Cutlip, R. C., et al.: An epizootic of Tyzzer's disease in rabbits. Lab. Anim. Sci., *21*:356-361, 1971.

Ganaway, J. R., Allen, A. M., and Moore, T. D.: Tyzzer's disease. Am. J. Pathol., *64*:717-732, 1971.

Niven, J. S. F.: Tyzzer's disease in laboratory animals. Z. Versuchstierkd., *10*:168-175, 1968.

Strittmatter, J.: Elimination of Tyzzer's disease in the Mongolian gerbil (*Meriones unguiculatus*) by fostering to mice. Z. Versuchstierkd., *14*:209-214, 1972.

Takaski, Y., Oghiso, Y., and Sato, K.: Tyzzer's disease in hamsters. Jap. J. Exp. Med., *44*:267-270, 1974.

Van Kruiningen, H. J., and Blodgett, S. B.: Tyzzer's disease in a Connecticut rabbitry. J. Am. Vet. Med. Assoc., *158*:1205-1212, 1971.

Bordetella bronchiseptica Infection

HOSTS

Clinical infection with *Bordetella bronchiseptica* is common in the guinea pig. Rats, rabbits, dogs, cats, and primates may also develop clinical infections, but these hosts are generally considered possible carriers. The infection can have disastrous consequences in a guinea pig colony.

ETIOLOGY

Bordetella bronchiseptica, a small, Gram-negative bacillus, is the causative organism. At 48 hours *Bordetella* colonies are 1 to 2 mm in diameter, yellowish-brown, and variably beta-hemolytic. Carbohydrates are not fermented.

TRANSMISSION

Transmission of *B. bronchiseptica* is by direct contact with clinically affected animals, carrier hosts, contaminated fomites, and respiratory aerosol.

PREDISPOSING FACTORS

Guinea pigs are highly susceptible to *B. bronchiseptica* infection, especially the young, the pregnant, or animals with subclinical scorbutus. Contact with carrier species, especially rabbits, must be avoided.

CLINICAL SIGNS

Clinical signs of *B. bronchiseptica* infection in guinea pigs vary from no signs to anorexia, reluctance to move, nasal discharge, dyspnea, and death. Abortions and stillbirths have been associated with infections in guinea pigs. *B. bronchiseptica* causes pneumonia and upper respiratory disease in rats. Rabbits frequently harbor *B. bronchiseptica* in their respiratory passages in a benign, carrier state or as the causative agent of rhinitis.

NECROPSY SIGNS

Mucopurulent rhinitis, tracheitis, and pulmonary consolidation are lesions of *Bordetella* infection in guinea pigs. Uterine infections have been reported. *Bordetella bronchiseptica* infection in rats may complicate pulmonary mycoplasmosis.

DIAGNOSIS

A definitive diagnosis is based on clinical signs and culture of *B. bronchiseptica* from respiratory tissues.

TREATMENT

Treatment of *Bordetella* pneumonia in guinea pigs is usually not practical except in individual pets. Sulfamethazine in the water (0.5 gm/ml) for 2 weeks may suppress, but rarely cure, an active infection. Chloramphenicol (30 mg/kg for 5 days) or cephaloridine (20 mg/kg for 5 days) may be used to treat bacterial infections in research animals. Penicillins should not be given to guinea pigs.

PREVENTION

Good husbandry, clean stock, and separation of carrier animals from guinea pigs are essential. A *Bordetella*-free colony must be managed on the closed-colony basis with entry restricted to guinea pigs known to be free of the organism.

A formalin-killed bacterin administered intramuscularly with incomplete Freund's adjuvant to guinea pigs results in a protective titer lasting approximately 4 to 6 months (Ganaway et al.).

PUBLIC HEALTH SIGNIFICANCE

The public health significance of a *B. bronchiseptica* infection in a research animal colony is minimal.

BIBLIOGRAPHY

Ganaway, J. R., Allen, A. M., and McPherson, C. W.: Prevention of acute *Bordetella bronchiseptica* pneumonia in a guinea pig colony. Lab. Anim. Care, *15*:156-162, 1965.

Nikkels, R. J., and Mullink, J. W. M. A.: *Bordetella bronchiseptica* pneumonia in guinea pigs. Description of the disease and elimination by vaccination. Z. Versuchstierkd., *13*:105-111, 1971.

Simpson, W., and Simmons, D. J. C.: Problems associated with the identification of *Bordetella bronchiseptica*. Lab. Anim., *10*:47-48, 1976.

Woode, G. N., and McLeod, N.: Control of acute *Bordetella bronchiseptica* pneumonia in a guinea pig colony. Lab. Anim., *1*:91-94, 1967.

Cestodiasis

HOSTS

Cestodes infect a wide range of species, including the common laboratory animals.

ETIOLOGY

Hymenolepis nana, the dwarf tapeworm. Adult forms are common in hamsters, rats, mice, and gerbils.

Hymenolepis diminuta, the rat tapeworm.

Taenia (*Cysticercus*) *pisiformis*. The larval form is common in the rabbit. Adults occur in carnivores.

Multiceps (*Coenurus*) *serialis*. The larval form is rarely found in rabbits. The adult form occurs in canines.

Taenia taeniaformis (*Cysticercus fasciolaris*). The larval form occurs in mice and rats. Adult forms occur in the intestines of felines and other carnivores.

TRANSMISSION

Hymenolepis nana

1. Ova are passed in the feces directly from the definitive host to another definitive host. Tissue migration is involved.
2. Ova are passed in the feces from the definitive host through beetles or fleas to another definitive host. No tissue migration is involved.
3. Autoinfection may occur, in which case ova mature in the intestinal lumen of the definitive host. Autoinfection commonly occurs in laboratory hamsters and mice.

Hymenolepis diminuta

The ova are shed in the feces of the definitive host and passed through insects to another definitive host (rat, mouse, hamster, for example).

Taenia pisiformis and
Multiceps serialis

The rabbit ingests embryonated ova passed in the feces of the definitive carnivore host.

Taenia taeniaformis

Rodents ingest embryonated eggs, and the embryos pass from the intestine to the liver. The life cycle can be completed in the laboratory.

PREDISPOSING FACTORS

General debility predisposes to the more serious effects of hymenolepid infection. The contamination of feed and bedding with the feces of infected carnivores facilitates the spread of cestodiasis in the colony.

CLINICAL SIGNS

Hymenolepis spp infection may be latent and inapparent or cause constipation, catarrhal diarrhea, chronic weight loss, and death.

Taenia pisiformis infection in the rabbit in most cases is benign; in heavy infections a chronic weight loss may occur.

Multiceps serialis infection in the rabbit results in cyst formation under the skin or in other tissues, including the muscles and brain. Clinical signs depend on the tissue affected and displaced.

Taenia taeniaformis infection in rodents has no clinical effect.

NECROPSY SIGNS

Hymenolepis spp infection may be asymptomatic or cause a catarrhal enteritis, emaciation, and, indirectly, abscessation of mesenteric lymph nodes. The adult worms are found in the small intestine and occasionally in the pancreatic and biliary ducts of rodents.

Taenia pisiformis cysts, containing a single scolex and measuring up to 2 cm in diameter, are found attached to the mesentery of the rabbit. The hepatic lesions are irregular white foci with sharp boundaries. These lesions, which result from the hepatic migration of the larvae, measure 2 to 3 mm in diameter but may coalesce. These foci grossly resemble the hepatic lesions of an *Eimeria stiedae* infection, but the *Eimeria* lesions usually have less distinct margins.

Multiceps serialis cysts, containing multiple scoleces and measuring up to 5 cm in diameter, occur most often in the subcutaneous tissues and the connective tissue of skeletal muscle in rabbits, but they may occur in other tissues.

Taenia taeniaformis larvae produce a few (under 10 in most cases) small (3 to 5 mm) cysts in the rodent liver. These cysts may be mistaken for hepatic abscesses.

DIAGNOSIS

Hymenolepis nana is identified by the oval ovum (44 to 62 × 30 to 55 μ) found in the feces or the adult tapeworm (20 to 30 mm × 1 mm) in the small intestine or pancreatic and biliary ducts. Onchospheres (16 to 25 × 24 to 30 mm) occur in the intestinal villae. *H. diminuta* adults measure 40 to 60 μ × 4 mm, and the ova 52 to 81 × 62 to 88 μ.

Taenia pisiformis larvae are recognized on necropsy in the liver as scattered white foci or in the abdominal cavity as cysts with a single scolex. The larval cysts are usually between 0.5 and 2.0 cm in diameter. Histologically the *Taenia* migration lesion in the hepatic parenchyma consists of focal granulomas containing scattered polymorphonuclear leukocytes. The *Eimeria* lesion involves the bile duct epithelium (destruction, hyperplasia, oocyst development) and adjacent portal tissue (fibrosis with chronic inflammatory cell infiltration).

Multiceps serialis larvae may be detected by observation, palpation, and aspiration of cyst fluid, or by demonstration of cysts on necrospy. *Taenia taeniaformis* larvae in rodents are detected in the liver on necropsy examination.

TREATMENT

Hymenolepis spp infection is treated with niclosamide given in a single oral dose at 10 mg/100 gm body weight and repeated 7 days later. *Taenia* spp infections are usually subclinical and are not treated. *Multiceps serialis* cysts can be removed surgically, unless they are numerous or inaccessible.

PREVENTION

Hymenolepis spp infections are prevented by the exclusion of infected animals and insect vermin from the colony. *H. diminuta* has an absolute requirement for an intermediate host, such as a cockroach.

Taenia and *Multiceps* infections are prevented by excluding the feces of carnivores from the rabbitry or rodent colony. Dogs and cats in the immediate area should be treated for tapeworms.

PUBLIC HEALTH SIGNIFICANCE

Hymenolepids are pathogenic for man and can cause enteric disease. *Taenia pisiformis* does not affect man, but *Taenia taeniaformis* may if the human eats a raw rodent. *Multiceps serialis* larvae can affect man as they affect other intermediate hosts.

BIBLIOGRAPHY

Balk, M. W., and Jones, S. R.: Hepatic cysticercosis in a mouse colony. J. Am. Vet. Med. Assoc., *157*:678-679, 1970.

Flatt, R. E., and Campbell, W. W.: Cysticercosis in rabbits: incidence and lesions of the naturally occurring disease in young domestic rabbits. Lab. Anim. Sci., *24*:914-918, 1974.

Flynn, R. J.: Parasites of Laboratory Animals. Ames, The Iowa State University Press, 1970, 155-202.

Leiper, R. T.:Specimens illustrating the larval development of *Hymenolepis* in the wall of the small intestine of the gerbillae. Trans. Roy. Soc. Trop. Med. Hyg., *26*:319-320, 1933.

Read, C. P.: *Hymenolepis diminuta* in the Syrian hamster. J. Parasitol., *73*:324, 1951.

Ronald, N. C., and Wagner, J. E.: Treatment of *Hymenolepis nana* in hamsters with yomesan (niclosamide). Lab. Anim. Sci., *25*:219-220, 1975.

Tucek, P. C., Woodard, J. C., and Moreland, A. F.: Fibrosarcoma associated with *Cysticercus fasciolaris*. Lab. Anim. Sci., *23*:401-407, 1973.

Worley, D. E.: Comparative studies on the migration and development of *Taenia pisiformis* larvae in laboratory rabbits. Lab. Anim. Sci., *24*:517-522, 1974.

Coccidiosis

HOSTS

Coccidial organisms affect a wide range of hosts, but *Eimeria* spp are host specific. Intestinal coccidiosis of rabbits is a major but self-limiting disease.

ETIOLOGY

Eimeria stiedae, the rabbit liver coccidium

Eimeria irresidua, of the rabbit small intestine

Eimeria magna, of the rabbit jejunum and ileum

Eimeria media, of the rabbit small and large intestine

Eimeria perforans, of the rabbit small intestine

(Mixed infections are common in rabbits.)

Eimeria caviae, the guinea pig coccidium

Eimeria falciformis, the mouse coccidium

Klossiella cobayae, the guinea pig kidney coccidium

Klossiella muris, the mouse kidney coccidium

TRANSMISSION

Transmission of *Eimeria* spp is by ingestion of sporulated oocysts. Oocysts are shed in the feces. Caprophagy does not allow time for sporulation; therefore, infection by this route is unlikely unless sporulation occurs in repeated passages through the gastrointestinal tract.

PREDISPOSING FACTORS

Eimeria spp often occur in the intestinal tract of clinically normal animals. In stressful conditions or following the ingestion of large infective doses of oocysts, disease may result. The condition is more common in weanling animals housed in unsanitary environments.

CLINICAL SIGNS

The weanling animal previously exposed to the aseptic and mildly antimicrobial maternal milk and lacking immunity against common pathogens is particularly susceptible to the proliferation of an aberrant intestinal flora. Although older animals may tolerate the pathogen and shed the organism over a long period, young animals often develop a fulminating, fatal disease.

Rabbits with hepatic coccidiosis usually exhibit no clinical signs; but in severe infections, particularly in the young, cachexia, abdominal distention, and death may occur.

Clinical signs associated with intestinal coccidiosis vary from inapparent to poor weight gains to an acute to chronic enteritis with diarrhea. The diarrhea seen with intestinal coccidiosis ranges from a soft, unformed stool to a watery, catarrhal, or bloody fluid. Acutely affected rabbits exhibit rapid weight loss, severe dehydration, polydipsia, and anorexia or ravenous appetite. Young animals infected with a large dose of a virulent coccidium (*E. irresidua* or *magna*) may die with no clinical signs other than sudden death.

NECROPSY SIGNS

In hepatic coccidiosis the liver is enlarged and contains irregularly shaped, slightly raised yellow-white foci, which, when incised, ooze a yellow exudate. The foci are often stringy in appearance, indicating the path of an affected bile duct through the hepatic parenchyma. The gall bladder and major extrahepatic ducts may be enlarged and contain a yellowish exudate. The number of foci and the degree of hepatomegaly are related to the number of infective oocysts and the resistance of the host.

The lesions of hepatic coccidiosis are confined to the intrahepatic and extrahepatic biliary system. Changes in the parenchyma are secondary to the bile duct hyperplasia and inflammation occurring in the portal areas.

Gross lesions seen with intestinal coccidiosis may be inapparent or include edema, reddening, and focal necrosis of the intestinal wall. Hemorrhage into the intestinal lumen occurs with *E. irresidua* and *magna*. The severity of the disease varies with the depth of schizont penetration into the lamina and submucosa and with the contribution of enteric organisms to the inflammatory process.

DIAGNOSIS

In heavy primary infections young rabbits may die before oocysts appear in the feces, but this outcome is not common. In these cases the demonstration of developmental coccidial stages in the intestinal mucosa is necessary.

In longer term cases, both with intestinal and hepatic coccidiosis, oocysts appear in the feces. The number of oocysts seen in a fecal sample reflects the developmental stage and not the severity of the primary infection. It is thus possible to have a case of fatal coccidiosis without oocysts appearing in the feces and oocysts appearing in the feces without clinical coccidiosis.

TREATMENT

Coccidiosis is best prevented; treatment often has little effect on the course of the infection. Even the efficacious sulfonamide coccidiostats inhibit only the asexual or schizont-merozoite stages. The treatment dose of sulfaquinoxaline is 3.5 gm powder per 4 L (1 gal) drinking water or 180 ml (6 oz) of a 20% stock sulfaquinoxaline solution per gallon water. Both combinations produce an approximate 0.1% solution. This solution is administered in the drinking water continuously for 2 weeks. Sulfaquinoxaline is also used to treat hepatic coccidiosis. Rabbits receiving sulfaquinoxaline should not be slaughtered for food within 10 days of treatment.

PREVENTION

As immunity develops following exposure to *Eimeria* spp, rabbits may become carriers. This factor, when combined with the difficulty of eliminating infective oocysts from bedding and soil, makes coccidiosis difficult to eliminate from a rabbit colony. Control of coccidiosis requires screening of resident and incoming animals for oocysts in the feces, culling of infected animals, and strict cleaning and disinfection procedures. Detergents and disinfectants and a good scrub brush remove the material in which viable oocysts persist, but killing oocysts is difficult. Hot 2% lye solution, 10% ammonia, methyl bromide gas, heat sterilization, and soil removal are means of eliminating oocysts. The use of all-wire hanging cages, water bottles or automatic waterers, and "J" type feeders is extremely important in preventing coccidiosis.

Rabbit feeds supplemented with 0.025% sulfaquinoxaline and fed continuously during the critical weaning period (3 to 8 wk) reduce coccidiosis and acute pasteurellosis in young rabbits. Sulfaquinoxaline is the rabbit food additive permitted by the Federal Food and Drug Administration, although animals treated should not be slaughtered for feed for 10 days posttreatment. Sulfaquinoxaline in many species is a vitamin K antagonist. The use of antimicrobial supplemented feeds is only a stopgap measure until the husbandry standards are improved.

PUBLIC HEALTH SIGNIFICANCE

Eimeria spp from laboratory animals do not infect man.

BIBLIOGRAPHY

Chapman, M. P.: The use of sulfaquinoxaline in the control of liver coccidiosis in domestic rabbits. Vet. Med., *43*:375-379, 1948.

Hoenig, V.: *Eimeria stiedae* infection in the rabbit: effect on bile flow and bromsulphthalein metabolism and elimination. Lab. Anim. Sci., *24*:66-71, 1974.

Horton-Smith, C., Taylor, E. L., and Turtle, E. E.: Ammonia fumigation for coccidial disinfection. Vet. Rec., 52:829-832, 1940.

Kleeberg, H. H., and Steeken, W.: Severe coccidiosis in guinea pigs. J. South Afr. Vet. Med. Assoc., *26*:49-52, 1963.

Long, P. L., and Burns-Brown, W.: The effect of methyl bromide on coccidial oocysts determined under controlled conditions. Vet. Rec., *90*:562-567, 1972.

McPherson, C. W., et al.: Eradication of coccidiosis from large breeding colony of rabbits. Proc. Anim. Care Panel, *12*:133-140, 1962.

Niilo, L.: Acquired resistance to reinfection of rabbits with *Eimeria magna*. Can. Vet. J., *8*:201-208, 1967.

Owen, D.: Life cycle of *Eimeria stiedae*. Nature (London), *27*:304, 1970.

Rutherford, R. L.: The life cycle of four intestinal coccidia of the domestic rabbit. J. Parasitol., *29*:10-32, 1943.

Weisbroth, S. H., and Scher, S.: Fatal intussusception associated with intestinal coccidiosis (*Eimeria perforans*) in a rabbit. Lab. Anim. Sci., *25*:79-81, 1975.

Corynebacterium kutscheri Infection

HOSTS

Mice and rats are the hosts.

ETIOLOGY

Corynebacterium kutscheri is a Gram-positive, diphtheroid rod. After 48-hour aerobic incubation on blood agar, colonies are circular, 1 to 4 mm in diameter, grey to yellow-white, smooth, and nonhemolytic.

TRANSMISSION

Transmission of *C. kutscheri* is fecal-oral and possibly by respiratory aerosol. The organism is an opportunistic pathogen uncommonly resident in the intestine and, possibly, the upper respiratory tract.

PREDISPOSING FACTORS

Corynebacterium kutscheri infections in mice and rats are usually, but not always, stress induced. Nutritional deficiencies, concomitant infections, and radiation exposure are predisposing stresses. Prolonged cortisone administration has precipitated a *C. kutscheri* outbreak in rats. Rats are more resistant to the spontaneous disease than mice.

CLINICAL SIGNS

The acute disease, with high morbidity and mortality, is characterized by emaciation, rapid respiration, nasal and ocular discharge, swollen joints, depression, and a hunched posture. In cutaneous infections the skin exhibits abscessation, ulceration, and fistulous tracts. Death usually occurs within a week.

The chronic infection, with low morbidity and mortality, may be inapparent or nonspecific.

NECROPSY SIGNS

The extension of the organism from the gut through the circulatory system results in focal embolic abscessation in a variety of organs, which may include the kidneys, liver, brain, heart, joints, lungs, and skin. The lesions are scattered, raised, caseopurulent foci up to 15 mm in diameter. Congestion and adhesions may be associated with visceral lesions.

DIAGNOSIS

A definitive diagnosis depends on the recovery of *C. kutscheri* from affected tissues.

TREATMENT

The rapid course of the clinical disease renders treatment of individual animals difficult. The bacterium is sensitive to a wide range of antibiotics, including chloramphenicol and tetracycline.

PREVENTION

Prevention of a *C. kutscheri* outbreak involves selection of clean stock, good husbandry practices, and the culling of affected animals.

PUBLIC HEALTH SIGNIFICANCE

Humans are not susceptible to *C. kutscheri* infection.

BIBLIOGRAPHY

Giddens, W. E., et al.: Pneumonia in rats due to infection with *Corynebacterium kutscheri*. Path. Vet., *5*:227-237, 1968.

Le Maistre, C., and Tompsett, R.: The emergence of pseudotuberculosis in rats given cortisone. J. Exp. Med., *95*:393-408, 1952.

Weisbroth, S. H., and Scher, S.: *Corynebacterium kutscheri* infection in the mouse. I. Report of an outbreak: Bacteriology and pathology of spontaneous infections. Lab. Anim. Care, *18*:451-458, 1968.

Dermatophytosis

HOSTS

Dermatophytes affect a wide range of hosts. Most reports of the condition in laboratory animals involve rabbits, guinea pigs, and rats.

ETIOLOGY

Trichophyton mentagrophytes is the agent usually encountered in dermatophytoses of guinea pigs and rabbits, *Microsporum* spp infections are more common in rats.

TRANSMISSION

Dermatophytes are transmitted by direct contact with spores on haircoats, bedding, and soil. Carrier animals with latent infections do occur.

PREDISPOSING FACTORS

Husbandry and stress factors which increase exposure or reduce resistance predispose to dermatophytoses. The young, aged, pregnant, and otherwise stressed are particularly susceptible.

CLINICAL SIGNS

Infections with the dermatophytes are often subclinical. Lesions of *T. mentagrophytes* in guinea pigs usually arise on the face and spread to the trunk and limbs. Lesions are irregular or ovoid-shaped, partially hairless, and have crusts or scabs over raised, pruritic areas. Lesions of *Microsporum* spp in rats are often diffuse and result in a generalized partial alopecia and slightly hyperemic, scaly skin.

DIAGNOSIS

A Wood's lamp (ultraviolet source) used in a darkened room reveals the yellow-green fluorescence of certain varieties of *Microsporum canis*. Epidermal scales and bedding debris may also appear as bright flecks and confuse the diagnosis.

A skin scraping from the margin of the lesion placed with a drop of 10% KOH and warmed provides a sample for microscopic examination. A drop of blue ink will stain the fungal structures. The solutions should stand for 5 minutes before examination for hyphae and arthrospores. Septate hyphae of both fungal species may be seen growing into hair shafts. Arthrospores (small, round structures) will be seen surrounding the hair shafts with both agents. When available, histologic sections stained with periodic acid-Schiff or silver stains will aid in confirming a diagnosis.

Dermatophyte culture media (Dermatophyte Test Media[1] or Fungassay[2]) inoculated with several broken hairs or a scraping or brushing from the margin of the lesion will turn red in 4 or 5 days if a dermatophyte is present. Certain saprophytic organisms will also convert the phenol red indicator. Because clinically inapparent dermatophyte infections are common, a vigorous brushing of the pelt with a toothbrush is a suggested screening method for obtaining an inoculum.

TREATMENT

Topical and systemic antifungal agents are available. Treatment of dermatophytoses should be undertaken only after consideration of the public health significance of the organisms involved and the prolonged therapeutic regimen required for a cure.

Topical antifungal creams are applied twice a day for a minimum of 4 weeks. Griseofulvin is administered at 25 mg/kg body weight daily in the water or feed (for guinea pigs this is approximately 0.8 gm/kg feed).

PREVENTION

Maintenance of high-level husbandry standards, particularly with the young, aged, pregnant, or otherwise stressed, is the important preventive measure.

PUBLIC HEALTH SIGNIFICANCE

Man, particularly children, can develop ringworm if exposed to dermatophytes on infected animals or contaminated bedding.

BIBLIOGRAPHY

Alteras, I., and Cojocaru, I.: Human infection by *Trichophyton mentagrophytes* from rabbits. Mykosen, *12*:543-544, 1969.

Banks, K. L., and Clarkson, T. B.: Naturally occurring dermatomycosis in the rabbit. J. Am. Vet. Med. Assoc., *151*:926-929, 1967.

Dolan, M. M., et al.: Ringworm epizootics in laboratory mice and rats: Experimental and accidental transmission of infection. J. Invest. Dermatol., *30*:23-25, 1958.

Fuentes, C. A., and Aboulafia, R.: *Trichophyton mentagrophytes* from apparently healthy guinea pigs. Arch. Dermatol., *71*:478-480, 1955.

Gentles, J. C.: Experimental ringworm in guinea pigs: Oral treatment with griseofulvin. Nature (London), *182*:476, 1958.

Hagen, K. W.: Ringworm in domestic rabbits: Oral treatment with griseofulvin. Lab. Anim. Care, *19*:635-638, 1969.

McAllister, H. A.: Dermatophytes and dermatophytoses. *In* Diagnostic Procedures in Veterinary Microbiology. 2nd Edition. Edited by G. R. Carter. Springfield, Ill., Charles C Thomas, 1973.

Pombier, E. C., and Kim, J. C. S.: An epizootic outbreak of ringworm in a guinea pig colony caused by *Trichophyton mentagrophytes*. Lab. Anim., *9*:215-221, 1975.

Smith, J. M. B.: Diseases of laboratory animals—mycotic. *In* C.R.C. Handbook of Laboratory Animal Science. Edited by E. C. Melby and N. H.

[1]Charles Pfizer Co., 235 East 42nd St., New York, N.Y. 10017.

[2]Pitman-Moore Inc., Washington Crossing, N.J. 08560.

Altman. Cleveland, Chemical Rubber Company Press, 1974, Vol II.
Vogel, R. A., and Timpe, A.: Spontaneous *Microsporum audouinii* infection in a guinea pig. J. Invest. Dermatol., *28*:311-312, 1957.
Weisbroth, S. H., and Scher, S.: *Microsporum gypseum* dermatophytosis in a rabbit. J. Am. Vet. Med. Assoc., *159*:629-634, 1971.
Young, C.: *Trichophyton mentagrophytes* infection of the Djungarian hamster (*Phodeopus sungorus*). Vet. Rec. *94*:287-289, 1974.

Ectromelia (Mouse Pox)

HOSTS

The mouse is the natural host of mouse pox. This potentially devastating mouse disease is rare in the United States but endemic in many areas of the world. Recent outbreaks in the United States have been traced to shipments from foreign countries.

ETIOLOGY

The mouse pox virus is a relatively large 150 × 250 mμ DNA pox virus in the vaccinia subgroup.

TRANSMISSION

The probable transmission routes are respiratory aerosol, contact with cutaneous crusts, and ingestion of contaminated feces. The virus is excreted from the intestinal tract for long periods following infection. Urine and ectoparasites may also disseminate the virus.

PREDISPOSING FACTORS

Latent infections may be converted to the clinical disease if the hosts are stressed. There are major differences in the responses of various mouse strains to the ectromelia virus. The AKR strain has no infection and no skin lesions; the BALB/c inbred mice develop minimal skin lesions but recover; DBA mice die with few cutaneous lesions; and A strain mice die with extensive skin lesions.

CLINICAL SIGNS

Mouse pox infections may be latent, acute, subacute, or chronic. The latent form may exist in the intestinal tract of older mice and not become apparent until the host is stressed.

The acute, systemic form of ectromelia with high morbidity and mortality occurs in epizootic outbreaks. Clinical signs of acute mouse pox include hunched posture, rough hair coat, conjunctivitis, facial swelling, and death. The cutaneous rash, an important source of virus dissemination, is seldom seen in acute outbreaks.

The subacute to chronic endemic or cutaneous form involves a generalized body rash, the swelling, necrosis, and sloughing of appendages, and sporadic deaths. The cutaneous lesions may resemble fight wounds or a *Streptobacillus moniliformis* infection.

NECROPSY SIGNS

The gross lesions associated with an acute ectromelia outbreak are hyperemia and edema of the visceral organs, enlargement of the spleen, a peritoneal exudate, and, as the disease progresses, hemorrhage into the intestinal lumen and focal necrosis of the liver, spleen, and pancreas.

In the subacute to chronic form, the focal necrosis becomes more extensive, the vesicular, crusted cutaneous pox lesions develop, and swelling and necrosis of the extremities occur.

DIAGNOSIS

Mouse pox is diagnosed by clinical and gross necropsy signs, by the demonstration of intracytoplasmic, eosinophilic inclusion bodies (Marchal bodies) in the epithelial cells of the skin, small intestine, and pancreas, by a fluorescent antibody test, and through the use of a hemagglutination inhibition test on sera from subacute or chronic cases. When inapparent infections are suspected, known susceptible, disease-free mice may be introduced into the colony as sentinel animals.

TREATMENT

There is no treatment for ectromelia. Vaccination may be used to control outbreaks; however, elimination of affected colonies is preferred to preclude the possibility of establishing endemic ectromelia.

PREVENTION

Selection of clean stock, careful husbandry and quarantine measures, serologic screening tests, and the use of susceptible sentinel mice are measures to exclude latent carriers from the colony. Particular care should be taken with strains imported from European colonies because enzootic mouse pox has existed in Europe. Imported mice or mice from unknown sources should be strictly quarantined. Likewise, mouse tissue products destined for transmission to susceptible mice should be checked for ectromelia virus by passage through susceptible hosts. Infected colonies must be eliminated, and rooms and equipment should be thoroughly decontaminated. The virus remains viable in dry environments.

Susceptible mice may be vaccinated with the IHD-T vaccinia strain, which does not produce HI antibodies. Such antibodies would interfere with subsequent serologic testing. The vaccine is given by scarification at the tail base. If a vaccination "take" is not obvious, a latent carrier state should be suspected. Successive passage of tumor lines through at least two vaccinated mice results in the loss of ectromelia virus from a tumor line.

PUBLIC HEALTH SIGNIFICANCE

Man may be susceptible to infection by the mouse pox virus.

BIBLIOGRAPHY

Briody, B. A.: Response of mice to ectromelia and vaccinia viruses. Bacteriol. Rev., *23*:61-95, 1959.

Briody, B. A.: The natural history of mouse pox. Natl. Cancer Inst. Monogr., *20*:105-116, 1966.

Fenner, F.: Mouse pox (infectious ectromelia of mice). J. Immunol., *63*:341-373, 1949.

Roberts, J. A.: Histopathogenesis of mouse pox. I. Respiratory infection. Br. J. Exp. Pathol., *43*:451-461, 1962.

Roberts, J. A.: Histopathogenesis of mouse pox. II. Cutaneous infection. Br. J. Exp. Pathol., *43*:462-468, 1962.

Trentin, J. J., and Ferrigno, M. A.: Control of mouse pox (infectious ectromelia) by immunization with vaccinia virus. J. Natl. Cancer Inst., *18*:757-767, 1967.

Whitney, R. A.: Ectromelia in U.S. mouse colonies: Letter. Science, *184*:609, 1974.

Encephalitozoonosis (Nosematosis)

HOSTS

Rabbits are the principal host among research animals, although mice, guinea pigs, and hamsters may be infected. The disease is common in rabbits.

ETIOLOGY

Encephalitozoon (*Nosema*) *cuniculi*, a microsporidium, is an obligate, intracellular, protozoan parasite.

TRANSMISSION

Urine-oral passage is the most important route of transmission in a rabbit colony. Fecal and transplacental routes are suspected. Ectoparasitic and endoparasitic vectors may be involved.

PREDISPOSING FACTORS

The severity of the infection, as with other host-parasite interactions, depends on host resistance and infective dose. Young rabbits in unsanitary conditions are at increased risk to develop the clinical disease.

CLINICAL SIGNS

Most cases of encephalitozoonosis are chronic and subclinical and therefore diagnosed only on postmortem examination. In rare cases, tremors, torticollis, paresis, convulsions, and death may occur. No signs of renal impairment have been reported.

NECROPSY SIGNS

Encephalitozoon has a predilection for the kidney and brain. The chronic renal infection is evidenced by numerous, randomly scattered, small (1 to 3 mm) pits on the cortical surface. In the earlier or more acute stages of the disease, the kidneys may be moderately enlarged. The central nervous system lesions are not grossly evident.

DIAGNOSIS

Encephalitozoon infection can be detected by an indirect fluorescent antibody test; by a skin test using an *Encephalitozoon* antigen; by the production of ascitic fluid (containing *E. cuniculi*) in susceptible mice following an intraperitoneal injection of infected tissue, or by histologic observation of characteristic focal granulomas and perivascular cuffing in the renal interstitium and brain. The organism, which stains well with Goodpasture's stain, appears as oval or crescent-shaped trophozoites 1 μm long or in pseudocysts 8 to 12 μm in diameter.

TREATMENT

No effective treatment has been reported.

PREVENTION

Selection of rabbits free of *E. cuniculi* is the best method to prevent colony infection. Urine and fecal transmission is reduced if sipper tube or automatic waterers and hopper feeders are used. Transplacental transmission complicates eradication measures.

PUBLIC HEALTH SIGNIFICANCE

A few cases of *Encephalitozoon* encephalitis have been reported in man from Russia and Japan.

BIBLIOGRAPHY

Flatt, R. E., and Jackson, S. J.: Renal nosematosis in young rabbits. Pathol. Vet., 7:492-497, 1970.

Hunt, R. D., King, N. W., and Foster, H. L.: Encephalitozoonosis: Evidence for vertical transmission. J. Infec. Dis., *126*:212-214, 1972.

Jackson, S. J., Solorzano, R. F., and Middleton, C. C.: An indirect fluorescent antibody test for antibodies to *Nosema cuniculi* in rabbits. Proc. 77th Ann. Meeting U.S. Anim. Hlth. Assoc., 1973, 478-490.

Koller, L. D.: Spontaneous *Nosema cuniculi* infection in laboratory rabbits. J. Am. Vet. Med. Assoc., *155*:1108-1114, 1969.

Lainson, R., et al.: Nosematosis, a microsporidial infection of rodents and other animals, including man. Br. Med. J., *2*:470-472, 1964.

Moller, T.: A survey on toxoplasmosis and encephalitozoonosis in laboratory animals. Z. Versuchstierkd., *10*:27-38, 1968.

Pakes, S. P., Shadduck, J. A., and Olsen, R. G.: A diagnostic skin test for encephalitozoonosis in rabbits. Lab. Anim. Sci., *22*:870-877, 1972.

Pattison, M., Clegg, F. G., and Duncan, A. L.: An outbreak of encephalomyelitis in broiler rabbits caused by Nosema cuniculi. Vet. Rec., *80*:404-405, 1971.

Ruge, H.: Encephalitozoon in the guinea pig. Zentralbl. Bakt., *156*:543-544, 1951.

Shadduck, J. A., and Pakes, S. P.: Encephalitozoonosis (nosematosis) and toxoplasmosis. Am. J. Pathol., *64*:657-674, 1971.

Epizootic Diarrhea of Infant Mice (EDIM)

HOSTS

Mice and perhaps rats are hosts for the EDIM virus.

ETIOLOGY

The EDIM virus contains RNA and is structurally and chemically related to a group of viruses resembling reoviruses. EDIM therefore has been tentatively classified in the Diplorna virus group.

TRANSMISSION

The EDIM virus is transmitted by direct contact, ingestion of contaminated feces, or by contaminated dusts and aerosols. The disease is highly contagious. Animals surviving the clinical disease or exposed as adults may become chronic shedders of the virus for days or weeks. The latent carrier condition is particularly important in breeding females.

PREDISPOSING FACTORS

The clinical diarrheal disease is seen only in mice between 7 and 21 days of age. EDIM is more severe in mice born to primiparous or nonimmune females. Carrier animals in the colony remain a source of infection. Strains vary in susceptibility to infection. A low ambient humidity (below 50%) may predispose to EDIM infection.

CLINICAL SIGNS

EDIM is a diarrheal disease of high morbidity and low mortality. Affected suckling mice have a soft, stringy, yellow feces that wets and stains the perineum. The dried feces accumulate in the rectum and around the anus to produce obstipation, which is responsible for the low mortality rate associated with the disease. Although affected mice continue to eat, they fail to grow normally, and stunted growth is a common sequella.

NECROPSY SIGNS

The gastrointestinal tract is distended with gas and a watery, yellow feces. Rectal impaction with a dried fecal mass may occur.

DIAGNOSIS

Clinical signs, a serum neutralization test, and an immunofluorescent test for viral antigen in the cytoplasm of the intestinal mucosal cells are diagnostic tests for EDIM. Intranuclear and intracytoplasmic inclusion bodies have been reported in the epithelial cells of the liver and intestine, although these inclusions may not be related to EDIM.

TREATMENT

The only treatment regimen to save the life of an affected mouse would be to periodically remove obstructing fecal masses from the anus. Obviously, this is not used with large numbers of mice.

PREVENTION

Selection of clean stock and the use of cage filter covers will reduce or prevent outbreaks of EDIM. Following an outbreak, the colony should be destroyed, and equipment should be cleaned and sterilized. Humidity in the room should be kept above 50% saturation.

PUBLIC HEALTH SIGNIFICANCE

The EDIM virus is not known to affect man.

BIBLIOGRAPHY

Banfield, W. G., Kasnic, G., and Blackwell, J. H.: Further observations on the virus of epizootic diarrhea of infant mice: An electron microscope study. Virology, *36*:411-421, 1968.

Cheever, F. S.: Epidemic diarrheal disease of suckling mice. Ann. N. Y. Acad. Sci., *66*:196, 1956.

Kraft, L. M.: Studies on the etiology and transmission of epidemic diarrhea of infant mice. J. Exp. Med., *106*:743-755, 1957.

Kraft, L. M., et al.: Practical control of diarrheal disease in a commercial mouse colony. Lab. Anim. Care, *14*:16-19, 1964.

Much, D. H., and Zajac, I.: Purification and characterization of epizootic diarrhea of infant mice virus. Infect. Immun., *6*:1019-1024, 1972.

Poiley, S. M.: The development of an effective method for control of epizootic diarrhea in infant mice. Lab. Anim. Care, *17*:501-510, 1967.

Heat Stroke

HOSTS

All animals are susceptible to heat stroke, but rabbits and guinea pigs are particularly susceptible.

ETIOLOGY

In high ambient temperatures, in combination with the predisposing factors listed, thermoregulatory mechanisms fail, and the body temperature rises beyond a level compatible with life.

PREDISPOSING FACTORS

Predisposing factors to heat stroke include an ambient temperature above 28° C (85° F), high humidity (above 70%), a thick hair coat, obesity, direct sunlight, poor ventilation, insufficient water, and crowding.

CLINICAL SIGNS

Signs of heat stress include hyperemia of peripheral vessels, rapid respiration, cyanosis, prostration, and death. Rodents have large salivary glands and respond to overheating by profuse salivation. Excess saliva escaping from the corners of the mouth to wet the body has a cooling effect through evaporation.

NECROPSY SIGNS

Lesions of heat stroke include congestion of the tissues, particularly of the lungs and intestinal wall.

DIAGNOSIS

The history of acute onset in hot weather and the absence of evidence of ketosis or infectious or toxic disease support a diagnosis of heat stroke.

TREATMENT

Heat-stressed animals can be sprayed with water, carefully dipped into a cool bath, or wrapped in a dampened cloth.

PREVENTION

Prevention of heat stress includes provision of shade, adequate air circulation, feed and water, water sprays to cool cages, or a container of ice. Obese and heavily furred animals are particularly susceptible and should be conditioned or eliminated.

BIBLIOGRAPHY

Daily, W. M., and Harrison, T. R.: A study of the mechanism and treatment of experimental heat pyrexia. Am. J. Med. Sci., *215*:42-54, 1948.

Jackson, R. F.: Heat stress (heat stroke, hyperpyrexia). *In* Current Veterinary Therapy V. Edited by R. W. Kirk. Philadelphia, W. B. Saunders, 1974.

Hypovitaminosis C

HOSTS

Guinea pigs and primates are the research animal groups with an absolute requirement for dietary ascorbic acid.

ETIOLOGY

Absence or deficiency in the guinea pig and primate of the hepatic enzyme 1-gulonolactone oxidase, which is necessary for the production of ascorbic acid from glucose, necessitates an exogenous or dietary supply of vitamin C. Ascorbic acid is involved in several biochemical processes, including the synthesis of collagen and the intercellular cement substance.

PREDISPOSING FACTORS

Age, sex, type of diet, pregnancy, lactation, disease status, and environmental conditions all affect the duration of onset and the signs of an ascorbic acid deficiency; however, all guinea pigs will show some deficiency symptoms within 2 weeks if vitamin C is withheld.

CLINICAL SIGNS

Weakness, anorexia, diarrhea, cutaneous sores, limb stiffness, reluctance to move, and death in 3 to 4 weeks from starvation or secondary infections are signs of vitamin C deficiency. Animals marginally deficient in vitamin C are highly susceptible to a wide variety of diseases, including infectious conditions.

In the young with subacute or chronic scorbutus, the costochondral junctions may become enlarged.

NECROPSY SIGNS

Gross lesions of scorbutus include widespread hemorrhages in muscles and periosteum, particularly around the stifle joint and rib cage, and deformity of the epiphyses in young guinea pigs. Epiphyseal enlargement is best noted at the costochondral junction.

DIAGNOSIS

The history and clinical and necropsy signs, particularly stiffness and hemorrhage around the stifle joints, provide a diagnosis. Secondary bacterial or metabolic diseases may obscure a primary, subclinical vitamin C deficiency.

TREATMENT

Ascorbic acid supplied daily via feed or water at 10 mg/kg body weight for maintenance or 30 mg/kg body weight during pregnancy will stop or reverse consequences of a deficiency. However, abrupt changes in the taste of the water may cause the guinea pig to refuse to drink. Oral ascorbic acid drops will then become necessary.

PREVENTION

Provision of adequate, fresh, stabilized vitamin C in feed or water at recommended daily levels (10 to 30 mg/kg body weight or 200 mg/L in the drinking water) will prevent hypovitaminosis C. One hundred gm fresh cabbage contains approximately 60 mg vitamin C. Approximately 120 ml or one-half cup of kale weighs about 50 gm and contains 60 mg of vitamin C. Brass, metal, hard water, and heat cause accelerated deterioration of ascorbic acid in solution. In water in an open crock up to 50% of the vitamin C activity is lost within 24 hours.

BIBLIOGRAPHY

Eva, J. K., Fifield, R., and Rickett, M.: Decomposition of supplementary vitamin C in diets compounded for laboratory animals. Lab. Anim., *10*:157-159, 1976.

Fullmer, H. M., Martin, G. R., and Burns, J. J.: Role of ascorbic acid in the formation and maintenance of dental structures. Ann. N.Y. Acad. Sci., 92:286-295, 1961.

Gershoff, S. N.: Nutrient Requirements of Laboratory Animals, 2nd revised ed. NAS/NRC, 1972.

Gore, I., Fujinami, T., and Shirahams, T.: Endothelial changes produced by ascorbic acid deficiency in guinea pigs. Arch. Pathol. (Chicago), *80*:371-376, 1965.

LaDu, B. N., and Zannoni, V. C.: The role of ascorbic acid in tyrosine metabolism. Ann. N.Y. Acad. Sci., *92*:175-192, 1961.

Navia, J. M., and Hunt, C. E.: Nutrition, nutritional diseases, and nutrition research. *In* The Biology of the Guinea Pig. Edited by J. E. Wagner and P. J. Manning. New York, Academic Press, 1976.

Pirani, C. L., Bly, C. G., and Sutherland, K.: Scorbutic arthropathy in the guinea pig. Arch. Pathol., *49*:710-732, 1950.

Reid, M. E.: Guinea pig nutrition. Proc. Anim. Care Panel, *8*:23-33, 1957.

Smith, D. F., and Balagura, S.: Taste and physiological need in vitamin C intake by guinea pigs. Physiol. Behav., *14*:545-549, 1975.

Lymphocytic Choriomeningitis

HOSTS

The natural reservoir host for the lymphocytic choriomeningitis (LCM) virus is the wild rodent population, where the prevalence of exposure and infection may reach 100%. The disease occurs naturally in mice, guinea pigs, chinchillas, canines, and primates and can be transmitted to hamsters, which have been the most recent source of multiple cases of human disease.

ETIOLOGY

The LCM virus is classified as an Arenavirus (RNA). Based on clinical effects, viscerotropic and neurotropic strains exist, although their effects are, in many reports, more quantitative than qualitative.

TRANSMISSION

Animals infected in utero or by ingestion or wound entry during the first few days post partum develop a subclinical persistent tolerant infection (PTI) in which viruses are shed in the urine and saliva. In the PTI no antibody titer can be detected by common serologic methods.

The virus may be transmitted in urine and saliva to susceptible hosts via traumatized skin, the conjunctiva, or the respiratory passages. Arthropod vectors (ticks, lice, mosquitoes) and dust may be vehicles for transmission.

PREDISPOSING FACTORS

Undetected carriers, decreased host resistance, and cutaneous traumatization favor the establishment of LCM in a colony. *Eperythrozoon coccoides* infection predisposes to the clinical disease.

CLINICAL SIGNS

The natural infection in rodents is usually asymptomatic. In the rare acute form, LCM is a fatal meningitis of young mice and other hosts, including man. Signs of infection include decreased growth, reluctance to move, limb extension, conjunctivitis, photophobia, and convulsions. Both transient and lifelong PTI infections exist. The progressive glomerulonephrosis associated with the deposition of antigen-antibody complex in PTI mice may cause weight loss in older animals.

NECROPSY SIGNS

Gross lesions in mice may include pleural exudates, visceral necrosis, pale liver, and splenomegaly. Lymphocytic infiltration of the meninges variably occurs. Cerebellar hypoplasia and retinopathy have been reported in rats.

DIAGNOSIS

Except for the PTI, infected animals either die or develop circulating antibodies, which can be detected by complement fixation or neutralizing antibody techniques. PTI infected animals may develop a glomerulonephrosis.

Injection of a tissue suspension from a suspected LCM infected animal into the foot pad of a susceptible mouse will cause an edematous reaction of the pad within 5 to 9 days. If tissue suspensions from a known LCM positive mouse are injected intracerebrally into an LCM-free mouse, the exposed mouse will develop neural signs, particularly if the mouse is suspended by the tail. If the injected mouse has had previous exposure to LCM, no disease will develop.

Intracerebral inoculation of a sterile, inert, protein-containing material (nutrient broth) will precipitate tremors in PTI mice. If these mice are held by the tail, they may convulse or spin. Guinea pigs injected intracranially with suspect tissue will die within 2 weeks if the inoculum contains LCM virus.

PREVENTION

The scanning diagnostic tests are methods for monitoring and maintaining an LCM-free colony. Filter cage covers reduce aerosol transmission, and the exclusion of insect and wild rodent vectors from the colony prevents introduction of the virus. Vertical, transuterine transmission of the LCM virus complicates eradication. No successful vaccine has been developed.

PUBLIC HEALTH SIGNIFICANCE

Lymphocytic choriomeningitis may be transmitted to man from infected animals. Signs in man may be (1) an influenza-like syndrome, (2) a meningeal infection, (3) a rarely fatal meningoencephalitis, or (4) an asymptomatic infection.

BIBLIOGRAPHY

Armstrong, D., et al.: Meningitis due to lymphocytic choriomeningitis virus endemic in a hamster colony. J. Am. Med. Assoc., *209*:265-266, 1969.
Hirsch, M. S., et al.: Lymphocytic choriomeningitis virus infection traced to a pet hamster. N. Engl. J. Med., *291*:610-612, 1974.
Hotchin, J.: Lymphocytic choriomeningitis resistant and slow virus infections. Monogr. Virol., *3*:2-70, 1971.
Hotchin, J.: The contamination of laboratory animals with lymphocytic choriomeningitis virus. Am. J. Pathol., *64*:747-769, 1971.
Maurer, F. D.: Lymphocytic choriomeningitis. J. Natl. Cancer Inst., *20*:867-870, 1958.
Smadel, J. E., and Wall, M. J.: Lymphocytic choriomeningitis in the Syrian hamster. J. Exp. Med., *75*:581-591, 1942.
Wilsnack, R. E.: Lymphocytic choriomeningitis. Natl. Cancer Inst. Monogr., *20*:77-84, 1966.

Mouse Encephalomyelitis (Mouse Polio)

HOST

Mice exhibit the clinical disease. Other rodents may develop antibody titers.

ETIOLOGY

The mouse encephalomyelitis virus is a small picornavirus (RNA). Existing serotypes, which vary in pathogenicity, include the TO, FA, and GD VII strains. GD VII has a group specific antigen.

TRANSMISSIONS

The mouse encephalomyelitis virus is a common resident of the intestinal tract of clinically normal mice. The virus is transmitted to weanling mice through the ingestion of contaminated feces.

PREDISPOSING FACTORS

The virus is a common, latent inhabitant of the mouse intestinal tract. Clinical disease may be precipitated if the host is stressed. Resistance to infection increases with age.

CLINICAL SIGNS

The infection occurs in preweanlings and, in the presence of serum antibodies, persists into adulthood. In the uncommon clinical disease, ruffled fur, huddling, circling, and flaccid paralysis occur. The FA and GD VII strains have shorter incubation periods and produce fewer paralytic signs than the TO serotypes.

NECROPSY SIGNS

Atrophy of skeletal muscle has been reported, but usually there are no gross lesions.

DIAGNOSIS

Diagnosis of mouse encephalomyelitis virus infection is based on a positive hemagglutination-inhibition test (using human type O cells), on the histologic demonstration of the necrosis of the ventral horn cells (TO strain), or on neuronal necrosis, gliosis, and perivascular cuffing (FA and GD VII) in the cord. Intracerebral inoculation of antibiotic-treated, virus-positive feces into 5-day-old mice will produce a flaccid paralysis or CNS lesions in 5 to 35 days.

TREATMENT

There is no treatment for mouse polio.

PREVENTION

The ubiquity of the enteric infection makes elimination difficult. A disease-free colony must be cesarean derived and maintained in a rigid, barrier-sustained system. A serologic screening test can be utilized to sample the colony.

PUBLIC HEALTH SIGNIFICANCE

Man is not infected by the mouse polio virus.

BIBLIOGRAPHY

Calisher, C. H., and Rowe, W. P.: Mouse hepatitis, Reo-3, and the Theiler viruses. *In* Viruses of Laboratory Rodents. Nat. Cancer Inst. Monogr., *20*:67-75, 1966.
Maurer, F. D.: Mouse poliomyelitis or Theiler's mouse encephalomyelitis. J. Nat. Cancer Inst., *20*:871-874, 1958.

Melmick, J. L. and Riordan, J. T.: Latent mouse encephalomyelitis. J. Immunol., *57*:331-342, 1947.
Theiler, M., and Gard, S.: Encephalomyelitis of mice. I. Characteristics and pathogenesis of the virus. J. Exp. Med., *72*:49-67, 1940.

Mouse Hepatitis

HOSTS

Mice are the hosts for the mouse hepatitis virus (MHV), although rats and hamsters may carry serum antibodies against this virus.

ETIOLOGY

The mouse hepatitis virus is a corona (RNA) virus. There are several strains in the MHV hepatoencephalitis group. Of the more common strains found in mice, all strains (MHV 1 through 4) may produce hepatitis, with MHV 4 more often causing encephalitis and MHV 3 an ascites.

TRANSMISSION

Mouse hepatitis virus infection is often a latent but always a highly contagious enteric infection. The virus is disseminated in the feces, by respiratory aerosol, or through cannibalism.

PREDISPOSING FACTORS

The mouse hepatitis virus, latent in the gut, will pass to other tissues if the host is debilitated, athymic, affected with certain lymphomas, injected with cortisone, or infected with the protozoan *Eperythrozoon coccoides*, all conditions which effect reticuloendothelial cell function. Older animals are more resistant, as are C3H and A strain mice. New Zealand Black mice are a susceptible strain. The louse *Polyplax serrata* may serve as a vector for *E. coccoides*.

CLINICAL SIGNS

Most MHV infections are latent, enzootic, and subclinical. Susceptible suckling mice exposed to infected adults may develop an encephalitis (tremors, spasticity) with high mortality. Other signs may include jaundice, a red-brown urine, and death.

Mouse hepatitis virus infections in susceptible suckling mice may occur alone or together with EDIM infections and produce a diarrheal disease syndrome (lethal intestinal virus of infant mice, LIVIM) with high mortality.

NECROPSY SIGNS

Gross lesions, if any, include grey or red foci on a yellowish liver. The kidneys may be swollen and congested.

DIAGNOSIS

A complement fixation test is commonly used to detect serum antibody to MHV infection. Female mice usually have higher titers than males.

TREATMENT

There is no treatment for MHV infection.

PREVENTION

Placement of cesarean-derived mice in a barrier-sustained colony, serologic sampling of the breeding females, and the use of filter-top cages are preventive measures. Ectoparasites should be eliminated.

PUBLIC HEALTH SIGNIFICANCE

There is no known public health significance with the mouse hepatitis virus.

BIBLIOGRAPHY

Cheever, F. S., et al.: A murine virus (JHM) causing disseminated encephalomyelitis with extensive destruction of myelin. J. Exp. Med., *90*:181-194, 1949.
Gledhill, A. W.: Enhancement of the pathogenicity of mouse hepatitis virus (MHIV) by prior infection of mice with certain leukemic agents. Br. J. Cancer, *15*:531-538, 1961.
Hartley, J. W., and Rowe, W. P.: Tissue culture

cytopathic and plaque assays for mouse hepatitis virus. Proc. Soc. Exp. Biol. Med., *113*:403-406, 1963.
ICLA Virus Reference Laboratory Sub-Committee: The Viruses of Laboratory Rodents and Lagomorphs. London, British Veterinary Association, 15-16, 1972.
Pollard, M., and Bussell, R. H.: Complement fixation with mouse hepatitis virus. Science, *126*:1245-1246, 1957.
Rowe, W. P., Hartley, J. W., and Capps, W. I.: Mouse hepatitis virus infection as a highly contagious, prevalent, enteric infection in mice. Proc. Soc. Exp. Biol. Med., *112*:161-165, 1963.

Mucoid Enteropathy

HOSTS

Rabbits are the hosts for mucoid enteropathy, a common enteric disease.

ETIOLOGY

Uncomplicated mucoid enteropathy, which is not primarily an inflammatory condition, is believed to result from an irritant, a toxin, or a stress-induced secretory alteration in the intestinal mucosa. Dietary changes, antibiotic influences, travel stresses, and numerous metabolic, pathogenic, and nutritional factors have been implicated as predisposing to or precipitating the disease.

Mucoid enteropathy may be complicated by inflammatory enteric conditions such as coccidiosis, colibacillosis, or Tyzzer's disease. Mucoid enteropathy has been described both as a distinct disease entity and as a sign associated with an enteric disease complex.

TRANSMISSION

If an infectious organism is involved, which is the case in some mucoid enteropathy outbreaks, transmission is fecal-oral.

PREDISPOSING FACTORS

Enteric disorders in rabbits are most common in weanling rabbits (3 to 10 wk), when there is a transition from the relatively sparse floral substrate of the neonatal gut to the established flora of the adult gut. If an aberrant flora becomes established during this critical transition, enteritis may result. Mucoid enteropathy is reported to be more common during the spring.

CLINICAL SIGNS

Abdominal distention, "slushing" of intestinal contents, a hunched posture, tooth grinding, rapid weight loss, dehydration, perineal staining, and death may be seen with mucoid enteropathy. Mucoid enteropathy is distinguished by the mucous diarrhea, subnormal temperature, and acute (1 to 2 week) course of the disease. Outbreaks may involve a high death rate.

NECROPSY SIGNS

Uncomplicated mucoid enteropathy is relatively noninflammatory; thus there is little hyperemia of the affected intestines. Except for bile staining, the primary enteropathy involves minimal organ discoloration. The stomach and small intestine contain gas and a watery fluid, and the large intestine has large accumulations of mucus. In cases with secondary protozoal or bacterial infection, hemorrhage, hyperemia, and edema are commonly seen in the intestinal wall.

DIAGNOSIS

Diagnosis of mucoid enteropathy is based on clinical and necropsy signs. Culture of the gut or a fecal examination may be indicated to detect colibacillosis, salmonellosis, or coccidiosis.

TREATMENT

Treatment is symptomatic and intended to prevent dehydration or reduce the proliferation of opportunistic pathogens. The treatment dose of sulfaquinoxaline for coccidiosis is 0.1% in the drinking water continuously for 2 weeks. Further suggested symptomatic treatments for diarrhea in research animals are mentioned in the discussion of nonspecific enteropathies. These treatments are untested, and the prognosis remains poor.

PREVENTION

Enteric conditions in rabbits are best prevented through the exercise of good management practices and not by treatment or routine feeding of antibiotic-supplemented feeds. Commercially available high fiber diets (15% to 25%) fed to rabbits may reduce the incidence of enteropathies in rabbits. Dimetridazole at 45 gm in 190 L (50 gal) drinking water administered during the weaning period (3 to 8 wk) has been shown to reduce the mortality during an outbreak.

PUBLIC HEALTH SIGNIFICANCE

There is no known public health significance for mucoid enteropathy.

BIBLIOGRAPHY

Casady, R. B., Hagen, K. W., and Sittmann, K.: Effect of high level antibiotic supplementation in the ration of growth and enteritis in young domestic rabbits. J. Anim. Sci., *23*:477-480, 1969.

Greenham, L. W.: Some preliminary observations on rabbit mucoid enteritis. Vet. Rec., *74*:79-85, 1962.

McCuistion, W. R.: Rabbit mucoid enteritis. Vet. Med. Small Anim. Clin., *59*:815-818, 1964.

Pout, D.: Mucoid enteritis in rabbits. Vet. Rec., *89*:214-216, 1971.

Van Kruiningen, H. J., and Williams, C. B.: Mucoid enteritis of rabbits. Comparison to cholera and cystic fibrosis. Vet. Pathol., *9*:53-77, 1972.

Vetesi, F., and Kutas, F.: Mucoid enteritis in the rabbit associated with *E. coli* changes in water, electrolyte and acid base balance. Acta Vet. Acad. Sci. Hung., *23*:381-388, 1973.

Whitney, J. C.: Treatment of enteric disease in the rabbit. Vet. Rec. *95*:533, 1974.

Murine Chronic Mycoplasmosis

HOSTS

Rats are the most common host of *Mycoplasma pulmonis* respiratory infection. Mice are also susceptible.

ETIOLOGY

Although *M. pulmonis* infection may overlie an existing viral infection or be complicated by a secondary bacterial infection, the pathogens primarily responsible for the clinical signs and lesions of murine chronic mycoplasmosis are one or more strains of *M. pulmonis*. The bacteria which may accompany *M. pulmonis* respiratory infection in murine rodents include *Pasteurella pneumotropica*, *Streptococcus pneumoniae*, *Bordetella bronchiseptica*, and *Corynebacterium kutscheri*. *Mycoplasma neurolyticum* may also produce respiratory disease in mice and rats.

TRANSMISSION

Transmission of *M. pulmonis* is by direct contact, placental passage, or short-distance respiratory aerosol.

PREDISPOSING FACTORS

Mycoplasma pulmonis is a ubiquitous organism transmitted to or present in the upper respiratory tract of the neonate. Most infections are subclinical. Crowding, ammonia, poor ventilation, and nonspecific stressors predispose to the development of the clinical disease.

CLINICAL SIGNS

Murine chronic respiratory disease of postweanling animals involves one or more sites in the respiratory tract and associated structures. Rhinitis is manifested as a serous or catarrhal nasal discharge and snuffling; conjunctivitis by an ocular discharge; otitis by a head tilt or scratching at the ears; laryngitis and tracheitis by chattering and coughing; and pneumonia by weight loss, labored breathing, and death. The nonspecific signs of a sick rodent (lethargy, hunched posture, and rough hair coat) are also seen. The disease may progress rapidly or linger for months.

NECROPSY SIGNS

The upper respiratory lesions involve serous to purulent inflammatory processes in affected tissues. Unilateral or bilateral otitis media is a common finding, often the only gross abnormality, in murine chronic mycoplasmosis. The pulmonary lesions in the early stages of the disease involve well-demarcated foci of firm red to grey consolidation. As the disease progresses, inflammatory debris accumulates in the air passages, resulting in bulging yellow foci of bronchiectasis. The content of these foci is a viscid to caseous yellow-grey material.

DIAGNOSIS

Diagnosis of murine mycoplasmosis is based on gross and microscopic lesions and on the isolation of *M. pulmonis* from nasal pharynx, tympanic bullae, uterus, trachea, or lungs. The organism may be carried in the upper respiratory passages in the absence of clinical disease. Culture of *M. pulmonis* requires special media (see Hayflick reference) enriched with yeast extract and 10% swine or horse serum. The plates are incubated at 37°C in an atmosphere of reduced oxygen and increased humidity.

TREATMENT

Elimination of a *M. pulmonis* infection from affected rats and mice is difficult. Antimicrobials placed in the drinking water for periods of a week or more will suppress *M. pulmonis* infections. Tetracycline at 2 to 5 mg/ml in the drinking water or sulfamerazine at 0.02% in the drinking water or 1 mg/4 gm feed is a suggested regimen for treating murine mycoplasmosis. Chloramphenicol may be injected intramuscularly at 30 mg/kg or Tylosin at 10 mg/kg for 5 days.

PREVENTION

Prevention of murine mycoplasmosis involves strict husbandry measures often impossible or unavailable in conventional rodent colonies. Cesarean-derived stock, rigid sanitation requirements, and a barrier-sustained colony operation are necessary to eliminate the disease. Disease-free rodents must be kept isolated from affected animals.

PUBLIC HEALTH SIGNIFICANCE

Mycoplasma pulmonis does not affect man.

BIBLIOGRAPHY

Brennan, P. C., et al.: Murine pneumonia: A review of the etiologic agents. Lab. Anim. Care, *19*:360-371, 1969.

Fallow, R. J.: Mycoplasmas and their role as rodent pathogens. Lab. Anim., *1*:43-53, 1967.

Ganaway, J. R., and Allen, A. M.: Chronic murine pneumonia in laboratory rats: production and description of pulmonary disease free rats. Lab. Anim. Care, *19*:71-79, 1969.

Greselin, E.: Detection of otitis media in the rat. Can. J. Comp. Med. Vet. Sci., *25*:274-276, 1961.

Habermann, R. T., et al.: The effect of orally administered sulfamerazine and chlortetracycline on chronic respiratory disease in rats. Lab. Anim. Care, *13*:28-40, 1963.

Hayflick, L.: Tissue cultures and mycoplasmas. Tex. Rep. Biol. Med., *23*:285-303, 1965.

Hill, A.: Transmission of *Mycoplasma pulmonis* between rats, Lab. Anim., *6*:331-336, 1972.

Innes, J. R. M., et al.: Establishment of a rat colony free from chronic murine pneumonia. Cornell Vet., *47*:260-280, 1957.

Kohn, D. F.: Sequential pathogenicity of *Mycoplasma pulmonis* in laboratory rats. Lab. Anim. Sci., *21*:849-855, 1971.

Lamb, D.: Rat lung pathology and quality of laboratory animals: the user's view. Lab. Anim., *9*:1-8, 1975.

Lindsey, R. J., and Cassell, G. H.: Experimental *Mycoplasma pulmonis* infection in pathogen-free mice: Models of studying mycoplasmosis of the respiratory tract. Am. J. Pathol., *72*:63-90, 1973.

Lindsey, J. R., et al.: Murine chronic respiratory disease: Significance as a research complication and experimental production with *Mycoplasma pulmonis*. Am. J. Pathol., *64*:675-680, 1971.

Neoplasia

Although literature surveys of the incidence of neoplasia in lagomorphs and rodents result in lengthy lists of several tumor types, each species, if not each strain, possesses a limited number of "common" tumors. Therefore, even though this discussion on neoplasia is limited to certain groups of tumors, it is important to note that the individual rabbit or rodent may develop one or more of a wide variety of neoplasms.

NEOPLASIA IN THE RABBIT

The uterine adenocarcinoma is the most common tumor of *Oryctolagus*, but its occurrence may be influenced by genetic background, age, and probably endocrinologic factors. This tumor has been reported more often in the Tan and Dutch breeds and less often in the Polish and Rex breeds. Rabbits under 3 years have an incidence of approximately 4%, whereas rabbits over 3 years have an incidence of uterine adenocarcinoma approaching 50% to 80%. There may also be higher incidences among hybrid than inbred strains. The association of uterine adenocarcinoma with pregnancy toxemia, pseudopregnancy, and hyperestrogenism remains controversial. The dose of estrogen may be the important factor determining its carcinogenic contribution.

If there is an immediate, predisposing influence, senile atrophy of the uterine endometrium probably assumes the role. With increasing age, the endometrial cells, beginning deep in the glandular crypts, become less specialized, and the connective tissue stroma becomes less cellular and more collagenous. These progressive, senile changes may underlie the transition to adenomatous and cystic hyperplasia and the *in situ* carcinoma. The position of cystic hyperplasia in the transition is uncertain.

The clinical signs associated with progressive uterine neoplasia in the rabbit include an altered reproductive performance and eventually death over a period of 5 to 20 months. During the subclinical, *in situ* period, affected does have a loss of fertility (even with unilateral involvement), reduced litter size, abortions, resorption, stillbirths, fetal retention, and a sanguineous vulvar discharge. The multiple neoplastic nodules, which may be from 1 to 5 cm in diameter by 6 months, can be palpated through the abdominal and uterine wall.

The hyperplastic stage lasts approximately 3 months, the carcinoma *in situ* until 7 months, and the metastatic stage from 10 to 12 months. The tumors are usually ovoid, firm, and hemorrhagic and are attached to the mesometrial junction and regularly spaced. The neoplasm may invade widely before metastasis occurs. Multiple tumors may be of different sizes, but they are usually at the same stage of differentiation.

The myxofibroma group of poxvirus-induced tumors occurs as fibromas in the *Sylvilagus* (cottontail) reservoir in both the Eastern and Western United States, Europe, South America, and Australia. These antigenically related viruses are transmitted mechanically by mosquitoes and biting insects from the cottontails to domestic rabbits, where disease processes range from peracute, fatal myxomatosis to transitory and single fibromas.

Myxomatosis is a highly fatal (mortality from 40% to 100%), estival, viremic disease transmitted from the California brush rabbit (*Sylvilagus bachmani*) to *Oryctolagus*. The virus induces a generalized proliferation of reticuloendothelial cells and a mucinous product, a process manifested clinically as massive subcutaneous swelling, particularly on the head, neck, and anogenital mucosa. Irregular, subcutaneous, gelatinous tumors accompany the "big head" syndrome. The development of the tumors is variable both in California and European outbreaks. Mortality varies with the strain of virus and resistance of the host; but death, when it occurs, usually occurs within 2 weeks of clinical onset. Attenuated myxomatosis and Shope fibroma origin vaccines have been prepared and are partially protective.

A related pox virus, endemic in Eastern cottontails (*S. floridanus*), will cause self-limiting, subcutaneous fibromas in adult, domestic rabbits. In young *Oryctolagus*, the disease may become a disseminated fibromatosis. Prevention of the myxofibromatous diseases involves the exclusion of mosquitoes from the rabbitry and vaccination during an epidemic.

Other neoplasms reported in the rabbit include lymphosarcoma, embryonal nephroma, papilloma, squamous cell carcinoma, bile duct tumor, osteogenic carcinoma, and others.

NEOPLASIA IN THE GUINEA PIG

Rhabdomyomatosis, an accumulation of glycogen-bearing myocardial cells, is visible grossly as pale foci on the mural and valvular endocardial surfaces of the atria and ventricles. The embryonic placentoma is a multiple germ layer transitiory product of parthenogenic development occurring within the ovary of the young female. The placentoma is resolved by fibrosis but may be related to the neoplastic ovarian teratoma.

True neoplastic processes are rare in guinea pigs, but an age-related increase in incidence has been noted. Estimates of the incidence of neoplasia in guinea pigs over 3 years of age range up to 30%. Pulmonary neoplasia, usually the bronchogenic papillary adenoma, is the most common category of tumors in the guinea pig, comprising 35% of the total. The second category (15%) is tumors of the skin and subcutis, although a single report of several trichofolliculomas biases the incidence. Tumors of the reproductive tract, the mammary glands, and the hematopoietic system comprise the remainder of the "common" guinea pig tumors. Spontaneous lymphocytic leukemia is an acute, virally induced (C-type RNA virus) lymphoblastic leukemia often fatal within 5 days of the onset of the leukemia. The affected animal's haircoat appears rough, the mucous membranes become pale, and the liver and lymphatic tissues are greatly enlarged. The white blood cell count may reach 250,000/mm^3.

NEOPLASIA IN THE HAMSTER

The incidence of spontaneous neoplasia in the golden hamster is low, but how low depends on the source of the survey and the ages of the animals surveyed. Reports of incidences vary from 4% as an overall population incidence to 50% or more in hamsters over 2 years of age. Although spontaneous neoplasia is uncommon in hamsters, these rodents are remarkably susceptible to a wide range of experimentally induced tumors. The hamster's eversible cheek pouch is an easily visualized, immunologically protected site for tumor transplantation.

As reports of incidences vary from colony to colony, an attempt at listing the most common hamster neoplasms may be fruitless, but tumors of the adrenal cortex (adenomas and adenocarcinomas) comprise the largest reported group. Next in incidence are tumors of the gastrointestinal tract (polyps, papillomas, and adenocarcinomas), tumors of the lymphoreticular system (lymphsarcoma, reticulum cell sarcoma), and tumors of the skin and subcutis (melanomas, fibromas, and carcinomas).

NEOPLASIA IN THE GERBIL

The incidence of spontaneous neoplasia in gerbils has been reported to be approximately 24%, with a higher incidence in aged animals. Pseudoadenomatous structures of the skin and cystic ovaries and periovarian cysts, which may cause infertility, are common, nonneoplastic processes occurring in the

Mongolian gerbil. Neoplasms in gerbils cover a wide range of types, with tumors of the female reproductive system perhaps most common. Granulosa cell tumors, uterine adenocarcinomas, lutein cell tumors, and dysgerminomas have been reported. Neoplasms of the skin, often found in association with the ventral scent gland, include basal cell carcinomas, sebaceous adenomas, and squamous cell carcinomas. Neoplasms have been reported in several other tissues, most being of mesenchymal origin.

NEOPLASIA IN THE MOUSE

Neoplasia in the mouse is one of the most extensively investigated disease processes in laboratory animals; volumes have been written about the causation, pathogenesis, and resolution of murine tumors. With the development of inbred strains and selection for tumor susceptibility and resistance the pattern of tumor occurrence has become quite different from the pattern in the undomesticated, wild house mouse. Mice develop a great variety of tumors, and, as in other animals, neoplasia may be manifested clinically as sudden death, weight gain or loss, infertility, cutaneous or subcutaneous swellings, and increased susceptibility to infectious disease.

Adult, "wild" female breeder mice have the following approximate tumor incidences: pulmonary tumors, 28%; hemangioendotheliomas, 8%; ovarian tumors, 6%; mammary tumors, 6%; hepatomas, 4%; leukemias, 2%, reticulum cell sarcomas, 2%; and subcutaneous sarcomas, 2%. Representative inbred strains and their approximate tumor incidences are described in the *Handbook on the Laboratory Mouse* by Charles G. Crispens (Springfield, IL, Charles C Thomas, Publisher, 1975).

Some representative inbred strains and their associated neoplasms are leukemias in AKR and C58 mice, mammary tumors in the C3H strain, pulmonary tumors in BALB/c, A, and SWR mice, and hepatomas in C3H and CBA mice. Tumor incidence varies with the sex, age, parity, and substrain of the mouse.

Oncogenic murine viruses (oncornaviruses) include the mouse mammary tumor virus (MTV), the mouse leukemia viruses (MLV), and the mouse sarcoma virus (MSV). The mouse mammary tumor virus, or the Bittner agent, is an RNA oncornavirus, which, when present in a susceptible strain, predisposes to mammary adenocarcinoma. The virus is widely distributed in the host and is passed to the fetus and, via the milk, to the neonate. The virus may be detected by electron microscopy or by the oncogenic consequences of injection into BALB/c mice.

The mouse leukemia viruses are RNA oncornaviruses. Transmission is primarily vertical, with passage through the placenta or milk. C58 and AKR strains are highly susceptible to leukemia. Other predisposing factors include radiation and chemical carcinogens.

NEOPLASIA IN THE RAT

As with the mouse and other research animals, the reported incidence of spontaneous neoplasia in the rat varies with the age, sex, strain, and environmental circumstances of the colony surveyed. Neoplasia in rats has a reported incidence up to 87% in rats over 2 years of age. The most common tumor in the rat is the mammary fibroadenoma, followed by uterine endometrial polyps, malignant lymphomas, testicular interstitial cell tumors, thyroid adenomas, pheochromocytomas, and pituitary adenomas.

Mammary neoplasms are most often benign fibroadenomas. These tumors, which may occur in males, are usually single and may grow to 8 or 10 cm in diameter. The fibroadenomas are well-demarcated, ovoid or discoid, firm, and nodular, although considerable variation exists in size, growth rate, color, and consistency. Mammary tumors of rats may be ulcerated, hemorrhagic, necrotic, or rarely malignant with metastasis. The benign tumor is well tolerated until the mass hinders locomotion or food consumption; rats with extremely large tumors lose weight and die. The encapsulated tumors may be surgically removed, although care should be taken to ligate the large vessels entering the mass. Since mammary tissue is widely distributed in the subcutis of the rat (and mouse), mammary tumors may be found behind the shoulders, on the ventral abdomen and flank, or around the tail base.

BIBLIOGRAPHY

Baba, N., and von Haam, E.: Animal model: Spontaneous adenocarcinoma in aged rabbits. Am. J. Pathol., *68*:653-656, 1972.

Benitz, K. F., and Kramer, A. W.: Spontaneous tumors in the Mongolian gerbil. Lab. Anim. Care, *15*:281-294, 1965.

Bullock, F. D., and Curtis, M. R.: Spontaneous tumors of the rat. J. Cancer Res., *14*:1-115, 1930.

Burrow, H.: Spontaneous uterine and mammary tumors in the rabbit. J. Pathol. Bacteriol., *51*:385-390, 1940.

Congdon, C. C., and Lorenz, E.: Leukemia in guinea pigs. Am. J. Pathol., *30*:337-351, 1954.

Crain, R. C.: Spontaneous tumors in the Rochester strain of the Wistar rat. Am. J. Pathol., *34*:311-315, 1958.

Crispens, C. G.: Handbook on the Laboratory Mouse. Springfield, Ill, Charles C Thomas, 1975.

Ediger, R. D., and Rabstein, M. M.: Spontaneous leukemia in a Hartley strain guinea pig. J. Am. Vet. Med. Assoc., *153*:954-956, 1968.

Fenner, F.: Classification of myxoma and fibroma viruses. Nature (London), *171*:562-563, 1953.

Fenner, F., and Woodroofe, T. M.: The pathogenesis of infectious myxomatosis: The mechanism of infection and the immunological response in the European rabbit (*Oryctolagus cuniculus*). Br. J. Exp. Pathol., *34*:400-411, 1953.

Flatt, R. E.: Pyometra and uterine adenocarcinoma in a rabbit. Lab. Anim. Care, *19*:398-401, 1969.

Fortner, J. G.: Spontaneous tumors, including gastrointestinal neoplasma and malignant melanomas in the Syrian hamster. Cancer, *10*:1153-1156, 1957.

Fox, R. R., et al.: Lymphosarcoma in the rabbit: Genetics and pathology. J. Natl. Cancer Inst., *45*:719-729, 1970.

Gilbert, C., and Gilman, J.: Spontaneous neoplasms in albino rats. S. Afr. J. Med. Sci., *23*:257-272, 1958.

Ginder, D. R.: Rabbit papillomas and the rabbit papilloma virus. A review. N.Y. Acad. Sci., *54*:1120-1125, 1952.

Green, H. S. N., and Strauss, J. S.: Multiple primary tumors in the rabbit. Cancer, *2*:673-691, 1949.

Hagen, K. W.: Spontaneous papillomatosis in domestic rabbits. Bull. Wildl. Dis. Assoc., *2*:108-110, 1966.

Hayden, D. W.: Generalized lymphosarcoma in a juvenile rabbit. A case report. Cornell Vet., *60*:73-82, 1970.

Heiman, J.: Spontaneous mammary carcinoma in a rabbit. Am. J. Cancer, *29*:93-101, 1937.

Joiner, G. N., Jardine, J. H., and Gleiser, C. A.: An epizootic of Shope fibromatosis in a commercial rabbitry. J. Am. Vet. Med. Assoc., *159*:1583-1587, 1971.

Kitchen, D. N., Carlton, W. W., and Bickford, A. A.: A report of fourteen spontaneous tumors of the guinea pig. Lab. Anim. Sci., *25*:92-102, 1975.

MacKenzie, W. F., and Garner, F. M.: Comparison of neoplasms in six sources of rats. J. Natl. Cancer Inst., *50*:1245-1257, 1973.

McKercher, D. G., and Saito, J.: An attenuated live virus vaccine for myxomatosis. Nature, *202*:933, 1964.

Marshall, I. D., Regnery, D. C., and Grodhaus, G.: Studies in the epidemiology of myxomatosis in California. I. Observation of two outbreaks of myxomatosis in coastal California and the recovery of myxoma virus from brush rabbit (*Sylvilagus bachmani*). Am. J. Hyg., *77*:195-204, 1963.

Newman, A. J., et al.: Spontaneous tumors of the central nervous system of laboratory rats. J. Comp. Pathol., *84*:39-50, 1974.

Noble, R. L., and Cutts, J. H.: Mammary tumors of the rat: A review. Cancer Res., *19*:1125-1139, 1959.

Prejean, J. D., et al.: Spontaneous tumors in Sprague-Dawley rats and Swiss mice. Cancer Res., *33*:2768-2773, 1973.

Pulley, L. T., and Shively, J. N.: Naturally occurring infectious fibroma in the domestic rabbit. Vet. Pathol., *10*:509-519, 1973.

Ringler, D. H., Lay, D. M., and Abrams, G. D.: Spontaneous neoplasms in aging Gerbillinae. Lab. Anim. Sci., *22*:407-414, 1972.

Rivers, T. M.: Infectious myxomatosis of rabbits. J. Exp. Med., *51*:965-979, 1930.

Roe, F. J. C.: Spontaneous tumors in rats and mice. Food Cosmet. Toxicol., *3*:707-720, 1965.

Rowe, W. P.: Genetic factors in the natural history of murine leukemia virus infection. Cancer Res., *33*:3061-3068, 1973.

Shumaker, R. C., et al.: Tumors in Gerbillinae: A literature review and a report of a case. Lab. Anim. Sci., *24*:688-690, 1974.

Stewart, F. W.: The fundamental pathology of infectious myxomatosis. Am. J. Cancer, *15*:2013-2028, 1931.

Szczech, G. M., et al.: Fibroma in Indiana cottontail

rabbits. J. Am. Vet. Med. Assoc., *165*:846-849, 1974.
Vink, H. H.: Rhabdomyomatosis (nodular glycogenic infiltration) of the heart in guinea pigs. J. Pathol., 97:331-334, 1969.
Weisbroth, S. H., and Scher, S.: Spontaneous oral papillomatosis in rabbits. J. Am. Vet. Med. Assoc., *157*:1940-1944, 1970.
Yabe, Y., et al.: Spontaneous tumors in hamsters: incidence, morphology, transplantation, and virus studies. Gann, *63*:329-336, 1972.

Nephrosis

HOSTS

Chronic renal disease occurs in several species, including rats, guinea pigs, and hamsters.

ETIOLOGY

The specific etiologies of chronic renal disease have not been fully elucidated. Chronic renal disease is a common, spontaneous, age, diet, and strain related, nonflammatory or mildly inflammatory disease of rats, guinea pigs, hamsters, and mice. Tubular occlusion by hypertrophy, or casts, periarteritis, or nonspecific microflora effects have been suggested as immediate causes of renal impairment in rats. A variable but large percentage of hamsters (to 88% in one report) over 12 months of age develop an interstitial or glomerular deposition of hyaline or amyloid. The lesion in aged guinea pigs is a nephrosclerosis. Lesions in mice are frequently due to autoimmune disease, particularly in the New Zealand black and white strains.

TRANSMISSION

These diseases are not presently believed to be contagious.

PREDISPOSING FACTORS

Advancing age and diets high in protein, carbohydrates, or calories, or low in potassium, may predispose to chronic renal disease in rats. Also, the severity or rate of progression of the disease varies with the strain of rat. Renal amyloidosis in hamsters and nephrosclerosis in guinea pigs occurs in otherwise healthy animals. Chronic bacterial infections have been related to an accelerated deposition of amyloid.

CLINICAL SIGNS

Chronic renal disease is usually a subclinical condition detected only on necropsy. In severe cases affected animals progressively lose weight, become depressed, and die. Polyuria may occur in hamsters. Mice with autoimmune nephrosis may show severe anemia and severe, generalized edema.

NECROPSY SIGNS

Chronic renal disease in rats is a chronic, bilateral nephrosis. In advanced cases the kidneys may be enlarged 2 to 3 times, discolored tan to yellow, granular or pitted, and have radial, pale striations on section. Cortical cysts may be up to 3 mm in diameter. Affected kidneys in other animals are enlarged, have irregularities on the cortical surface, and are pale. Mice may show evidence of edema and anemia.

DIAGNOSIS

Diagnosis of chronic nephrosis in rats is based on gross and microscopic lesions. Histologic abnormalities include dilation of the tubules (loop of Henle and distal convoluted tubules) with proteinaceous material, atrophy of the tubular epithelium, and fibrosis and mild lymphocytic infiltration of the renal interstitium. These lesions may be focal at the corticomedullary junction, or they may be wedge-shaped, with the broad portion at the capsule. The glomeruli and capsule, if affected, are thickened. Amyloid may be demonstrated either with a differential stain (Congo Red) or with polarized light.

TREATMENT

There is no specific treatment for chronic renal disease.

PREVENTION

Reducing the food intake of rats may slow the progression of the disease, but as the specific causative factors are unknown, prevention is not practiced.

PUBLIC HEALTH SIGNIFICANCE

There is no known public health significance.

BIBLIOGRAPHY

Andrew, W., and Pruett, D.: Senile changes in the kidney of Wistar Institute rats. Am. J. Anat., *100*:51-80, 1957.

Berg, B. N.: Spontaneous nephrosis with proteinuria, hyperglobinemia, and hypercholesterolemia in the rat. Proc. Soc. Exp. Biol. Med., *119*:417-420, 1965.

Blatherwick, N. R., and Medlar, E. M.: Chronic nephritis in rats fed high protein diets. Arch. Intern. Med., *59*:572-596, 1937.

Bras, G.: Age associated kidney lesions in the rat. J. Inf. Dis., *120*:131-134, 1969.

Bras, G., and Ross, M. H.: Kidney disease and nutrition in the rat. Toxicol. Appl. Pharmacol., *6*:247-262, 1964.

Gleiser, C. A., et al.: Amyloidosis and renal paramyloid in a closed hamster colony. Lab. Anim. Sci., *21*:197-202, 1971.

Heymann, W., and Lund, H. Z.: Nephrotic syndrome in rats. Pediatrics, *7*:691-706, 1951.

Snell, K. C.: Renal disease of the rat. *In* The Pathology of Laboratory Rats and Mice. Edited by E. Cotchin and F. J. C. Roe. Philadelphia, F. A. Davis, Co., 1967.

Pasteurella multocida Infection

HOSTS

Pasteurella multocida infection is a disease of major importance in rabbits, but guinea pigs and rats may rarely become infected.

ETIOLOGY

Pasteurella multocida, a small, Gram-negative, bipolar-staining, nonhemolytic, ovoid rod, commonly causes clinical disease in rabbits. There are several known serotypes of the organism.

TRANSMISSION

Pasteurella multocida is transmitted by respiratory aerosol and direct contact. The passage of the organism from chronically infected adults to susceptible offspring is an important consideration in preventing the disease.

PREDISPOSING FACTORS

Substandard husbandry and environmental stressors predispose to pasteurellosis, particularly in the young.

CLINICAL SIGNS

Clinical signs of *P. multocida* infection range from peracute, septicemic death to chronic abscessation or rhinitis. The primary locus of an infection can occur in almost any tissue and spread to any other tissue. Signs involved may include nasal discharge, snuffling, runny eyes, torticollis, cutaneous and subcutaneous abscesses, enlarged testes, vaginal discharge, infertility, and sudden death. Acute pneumonia is the common clinical syndrome in young rabbits. Pulmonary abscesses and anoxia may cause cyanosis of the iris, particularly apparent in albino rabbits.

NECROPSY SIGNS

Gross lesions of *P. multocida* infection may involve one or more of the following: (1) generalized visceral congestion and focal hemorrhage resulting from a septicemia, (2) well-demarcated reddish-grey foci of bronchopneumonia, and (3) fibrinopurulent or mucopurulent reactions in the meninges and brain, the middle ear, thoracic and abdominal viscera, the nasal passages, subcutaneous tissues, and the reproductive organs. Pus in a *P. multocida* abscess is creamy-white.

DIAGNOSIS

Diagnosis of a *P. multocida* infection is based on the isolation of the causative organism from blood or affected tissues. Purulent processes in rabbits are usually due to *P. multocida* or *Staphylococcus aureus*. Both grow on blood agar media. *Pasteurella multocida* is nonhemolytic and produces large, translucent, mucoid colonies up to 4 mm in diameter. *Staphylococcus* colonies are smaller, dry, opaque, and often hemolytic. Gram staining will provide a final diagnosis. *Pasteurella multocida* must also be differentiated from *Pasteurella pneumotropica* and *Bordetella bronchiseptica* through the use of indole, glucose, and urea cultures.

TREATMENT

Chronic pasteurellosis in rabbits involves purulent processes, and concerns in treatment include the number and location of abscesses, presence of fistulous tracts, the potential for septicemia if the abscess is ruptured, and the antibiotic sensitivity of the causative agent. If surgical drainage is indicated, the rabbit is restrained with ketamine (25 mg/kg IM). The abscess is opened, drained, and flushed three times a day for 5 days with chlorhexidine or tincture of iodine. With or without surgery, a systemic antibiotic is injected for approximately one week. *In vitro*, *P. multocida* is sensitive to ampicillin, cephaloglycin, chloramphenicol, erythromycin, gentamicin, kanamycin, polymyxin B, triple sulfa, and neomycin. Sensitivities may vary with bacterial strain.

Rhinitis is difficult to cure because the pathogenic organisms are sequestered within the nasal passage and isolated from systemic antibiotics. Aerosol therapy provides an approach to reaching and eliminating these organisms. Concerns in rhinitis treatment include the edematous membranes, copious secretions, antibiotic sensitivity of the bacteria, and host susceptibility to bronchospasm. For rhinitis, or "snuffles," a suggested but untested treatment is to expose the rabbit to a nebulized normal saline based spray delivering daily 200 mg kanamycin or neomycin and 0.5 mg isoetharine or isoproterenol (bronchodilators). Acetylcysteine at 1 ml of a 20% solution may also be added. Other drugs may be administered intranasally or systemically.

PREVENTION

Rabbit pasteurellosis is endemic in most rabbitries and difficult to eradicate. Selection of *Pasteurella*-free stock, repeated culture surveys of rabbits in the colony with culling or isolation of affected individuals, and provision for a 3 week entry quarantine period with cultural screening are methods of preventing and eliminating *P. multocida* from a colony.

High husbandry standards and the elimination of environmental stressors reduce the development of the more severe forms of the disease in adults. As the young are susceptible to pulmonary infection, chronically infected does should be culled. Weaning the young at 4 or 5 weeks reduces exposure to an infected mother.

Pasteurella multocida bacterins have not been effective in rabbits, but the prophylactic use of sulfaquinoxaline-supplemented rabbit feed may reduce the incidence of acute pneumonia in the exposed young. The use of sulfaquinoxaline, however, should be continued only until proper husbandry measures can be instituted.

PUBLIC HEALTH SIGNIFICANCE

Although humans can develop a cutaneous infection with *P. multocida*, the condition is not a concern to persons working with rabbits.

BIBLIOGRAPHY

Belin, R. P., and Banta, R. G.: Successful control of snuffles in a rabbit colony. J. Am. Vet. Med. Assoc., *159*:622-623, 1971.
Boisvert, P. L., and Fouser, M. D.: Human infection with *Pasteurella lepiseptica* following a rabbit bite. J. Am. Vet. Med. Assoc., *116*:1901-1902, 1948.
Campbell, A. M., and Lenz, J. L.: Treatment of subcutaneous abscesses with chlorhexidine. Synapse, *8*:6-7, 1975.
Flatt, R. E., and Dungworth, D. L.: Enzootic pneumonia in rabbits: Naturally occurring lesions in lungs of apparently healthy young rabbits. Am. J. Vet. Res., *32*:621-626, 1971.
Fox, R. R., Norber, R. F., and Myers, D. D.: The relationship of *Pasteurella multocida* to otitis media in the domestic rabbit (*Oryctolagus cuniculus*). Lab. Anim. Care, *21*:54-58, 1971.
Hagen, K. W.: Enzootic pasteurellosis in domestic rabbits. II. Strain types and methods of control. Lab. Anim. Care, *16*:487-491, 1966.
Hagen, K. W.: Chronic respiratory infection in the domestic rabbit. Proc. Anim. Care Panel, *9*:55-60, 1959.
Hagen, K. W.: Enzootic pasteurellosis in domestic rabbits. I. Pathology and bacteriology. J. Am. Vet. Med. Assoc., *133*:77-80, 1958.
Huebner, R. A.: *Pasteurella* pyometra in a rabbit. J. Am. Vet. Med. Assoc., *93*:389, 1938.
McKenna, J. M., South, F. E., and Musacchia, X.: *Pasteurella* infections in irradiated hamsters. Lab. Anim. Care, *20*:443-446, 1970.
Savage, N. C., and Sheldon, W. G.: Torticollis in mice due to *Pasteurella multocida* infection. Can. J. Comp. Med., *35*:267-268, 1971.
Wright, J.: An epidemic of *Pasteurella* infection in guinea pig stock. J. Pathol. Bacteriol., *42*:209-212, 1936.

Pasteurella pneumotropica Infection

HOSTS

Mice and occasionally rats and hamsters are infected by *Pasteurella pneumotropica*.

ETIOLOGY

Pasteurella pneumotropica is a Gram-negative, bipolar rod. Twenty-four hour colonies are surrounded by a zone of incomplete hemolysis (slight yellowish discoloration around the colonies).

TRANSMISSION

Pasteurella pneumotropica, which often exists in a latent, carrier state in the upper respiratory tract and gastrointestinal tract, may be spread by respiratory aerosol or fecal contamination.

PREDISPOSING FACTORS

As *P. pneumotropica* is often encountered as a latent, potential pathogen, stress conditions, particularly other infections, may precipitate the clinical disease. *Pasteurella pneumotropica* in conjunction with Sendai virus infection of mice frequently results in a fatal pulmonary syndrome. *P. pneumotropica* may complicate *Mycoplasma pulmonis* infections of rats and mice.

CLINICAL SIGNS

An infection with *P. pneumotropica* may be latent, secondary to another pathogen (Sendai virus in mice or *Mycoplasma pulmonis* in rats), or primary. Signs associated with *P. pneumotropica* infection include chattering, labored respiration, weight loss, cutaneous abscesses, conjunctivitis, panophthalmitis, mastitis, and harderian gland and subcutaneous abscesses in conventional and nude mice.

NECROPSY SIGNS

The pulmonary lesions in the early stages of *P. pneumotropica* pneumonia resemble the well-demarcated, red foci of consolidation seen with a *M. pulmonis* infection; however, the bacterial infection produces scattered abscessation. Suppurative reactions may also be encountered in the tympanic bullae, orbital glands, uterus, skin, mammary gland, cervical lymph nodes, and the urinary system.

DIAGNOSIS

A definitive diagnosis of *P. pneumotropica* infection is established through recovery of the organism on culture. On 24-hour incubation on blood agar the colonies are small (1 mm), circular, convex, smooth, and surrounded by a zone of slight greenish discoloration.

TREATMENT

Pasteurella pneumotropica, one of the common etiologic agents in murine respiratory disease, is sensitive to several antibiotics, including chloromycetin and ampicillin. Many isolants are resistant to tetracyclines and streptomycin.

PREVENTION

Elimination of murine respiratory infection from a colony requires a known disease-free stock placed into a clean and barrier-sustained colony. New animals should be quarantined until their carrier and disease status is known.

PUBLIC HEALTH SIGNIFICANCE

A strain of *P. pneumotropica* affects man, but the possibility of rodent-man transmission is unlikely.

BIBLIOGRAPHY

Brennan, P. C., Fritz, T. E., and Flynn, R. J.: *Pasteurella pneumotropica:* Cultural and biochemical characteristics, and its association with disease in laboratory animals. Lab. Anim. Care, *15*:307-312, 1965.

Brennan, P. C., Fritz, T. E., and Flynn, R. J.: Role of *Pasteurella pneumotropica* and *Mycoplasma pulmonis* in murine pneumonia. J. Bacteriol., 97:337-349, 1969.

Flynn, R. J., et al.: Urine infection in mice. Z. Versuchstierkd., *10*:131-136, 1968.

Moore, T. D., Allen, A. M., and Ganaway, J. R.: Latent *Pasteurella pneumotropica* infection of the gnotobiotic and barrier-held rats. Lab. Anim. Sci., 23:657-661, 1973.

Wagner, J. E., et al.: Spontaneous conjunctivitis and dacryoadenitis of mice. J. Am. Vet. Med. Assoc., *155*:1211-1217, 1969.

Weisbroth, S. H., Scher, S., and Boman, I.: *Pasteurella pneumotropica* abscess syndrome in a mouse colony. J. Am. Vet. Med. Assoc., *155*:1206-1210, 1969.

Pediculosis

HOSTS

Lice as a group have a wide host range, but individual species are host specific. Mites and lice sometimes occur together on a host.

ETIOLOGY

Haemodipsus ventricosis, the rabbit louse

Gliricola porcelli, the slender guinea pig louse

Gyropus ovalis, the oval guinea pig louse

Polyplax serrata, the mouse louse

Polyplax spinulosa, the spiny rat louse

Of the five species listed above, *Gliricola* and *Polyplax* spp are fairly common, and *Gyropus* and *Haemodipsus* are uncommon.

TRANSMISSION

Transmission of lice is by direct contact with an infected host or bedding. Lice seldom leave the host.

PREDISPOSING FACTORS

Young animals, or animals with decreased resistance housed under unsanitary conditions, may experience more severe infections and clinical consequences.

CLINICAL SIGNS

Infestation by the rabbit louse, *H. ventricosis*, may cause pruritus, alopecia, and anemia. The rabbit louse more often affects dorsolateral areas of the trunk.

Gliricola and *Gyropus* in heavy infestations may cause partial alopecia and scratching, usually around the ears.

Polyplax spp are blood-sucking lice and may cause debilitation, anemia, scratching, and death. *P. serrata* transmits the agent of murine eperythrozoonosis, and *P. spinulosa* the agent of murine haemobartonellosis.

DIAGNOSIS

Using a hand lens or dissecting microscope, observe the pelt, especially the margin of lesions and the nape of the neck, for adult or immature ectoparasites. If the animal is dead, direct observation of lice is easier if the pelt is cooled in the refrigerator for 30 minutes, removed for 10 minutes, and then examined with a lens. The parasites migrate from the cool skin toward the warmer hair tips. Placing the suspect pelt on a black paper and within a frame of double-gummed cellophane tape will facilitate detection of the lice.

Haemodipsus is 1.2 to 2.5 mm long and has an oval abdomen. *Polyplax* spp are slender and from 0.6 to 1.5 mm long. *Gliricola* has a narrow head and body approximately 1.0 to 1.5 mm long. *Gyropus* has a wide head and oval abdomen and is 1.0 to 1.2 mm long.

TREATMENT

Lice may be treated with the dusts and dips described in the section on acariasis. Resin strips containing dichlorvos placed on or near the cage for several 24-hour periods at one week intervals will effectively reduce ectoparasite populations in a colony. Elimination of ectoparasites from premises and equipment requires treatment and removal of the animals from the room and thorough mechanical scrubbing and formaldehyde fumigation of the premises.

PREVENTION

Pediculosis is prevented by installing clean stock into a clean facility. If lice are a problem, all animals should be treated, and the equipment should be cleaned and disinfected. Wild rodents and lagomorphs frequently have pediculosis and should be excluded from colonies of domestic animals.

PUBLIC HEALTH SIGNIFICANCE

Haemodipsus ventricosis is a vector for the transmission of *Francisella tularensis* from rabbits to man.

Gliricola and *Gyropus* are not known to affect man.

Polyplax spp may serve as vectors for rickettsial organisms.

BIBLIOGRAPHY

Eliot, C. P.: The insect vector for the natural transmission of *Eperythrozoon coccoides* in mice. Science, *84*:397, 1936.

Ferris, G. F.: The Suckling Lice. Mem. Pac. Coast Entomol. Soc., San Francisco, California.

Flynn, R. J.: Parasites of Laboratory Animals. Ames, The Iowa State University Press, 1973.

Murray, M. D.: The ecology of the louse *Polyplax serrata* (Burm.) on the mouse *Mus musculus* L. Aust. J. Zool., 9:1-13, 1961.

Pratt, H. D., and Karp, H.: Notes on the rat lice *Polyplax spinulosa* (Burmeister) and *Hoplopleura oenomydis* Ferris. J. Parasitol., 39:495-504, 1953.

Pregnancy Toxemia

HOSTS

Among the common laboratory animals, the guinea pig and rabbit are frequently affected with pregnancy toxemia, a metabolic disorder. Other animals that may have a similar condition are cattle, sheep, goats, horses, and man.

ETIOLOGY

The primary cause of a pregnancy toxemia probably involves uterine anoxia and an endocrine imbalance. Acidosis, ketosis, fatty liver, hypoglycemia, and hypocalcemia are secondary manifestations of the disease.

PREDISPOSING FACTORS

Factors predisposing a guinea pig to pregnancy toxemia are multiple. Specific predisposing influences include obesity, sudden dietary or nutritional changes, anorexia, displacement of gastric volume by increased fetal load, hereditary patterns, lack of exercise, sex, parity, and nonspecific environmental stressors. Guinea pigs exhibit a higher incidence of ketosis with obesity and later pregnancies and are more often affected in the winter, perhaps because of environmental or nutritional extremes or variations. Obese boars may also develop a similar toxemia.

Obesity and fasting are also predisposing factors in rabbits. The Dutch, Polish, and English breeds have an apparent predisposition to toxemia, as do multiparous does. Pregnant, pseudopregnant, and postparturient does are most often affected, but resting does and bucks may also develop the toxemia.

CLINICAL SIGNS

Pregnancy toxemia, a metabolic condition of low morbidity and high mortality, usually occurs during the last week of gestation in rabbits and during the last 2 weeks or first week post partum in guinea pigs. The condition may be asymptomatic, rapidly fatal, or involve depression, reluctance to move, incoordination, anorexia, dyspnea, convulsions, or abortion over a 1-to-5-day period. The urine becomes clear, and acetone breath, proteinuria, and ketonuria may develop. In the rabbit blood calcium decreases, blood phosphorus increases, nonprotein nitrogen increases, and blood glucose levels vary from hypoglycemic to hyperglycemic. Hematologic changes in the guinea pig are similar, except that hypoglycemia is a more frequent finding.

NECROPSY SIGNS

Gross signs of pregnancy toxemia include reduced gastric content, ample fat stores, near-term fetuses, uterine hemorrhage, and tan to yellow (fatty) liver and kidneys.

TREATMENT

If a toxemia is detected antemortem, administration of lactated Ringer's solution, calcium gluconate, or 5% dextrose, in conjunction with corticosteroids, could be tried as a treatment, but the prognosis is poor. Guinea pigs under treatment for ketosis frequently die from an acute enteritis apparently due, in part, to a stressor and lack of ingesta in the digestive tract.

PREVENTION

The critical consideration in preventing pregnancy toxemia is to supply an adequate amount of digestible energy during late pregnancy. Supplying a palatable high-energy feed such as calf starter pellets during late pregnancy may afford some protection. As obesity is an important predisposing factor, animals should be maintained in a good but nonobese condition prior to breeding.

BIBLIOGRAPHY

Abitol, M. M., et al.: Production of experimental toxemia in pregnant rabbits. Am. J. Obst. Gyne., *124*:460-470, 1976.

Foley, E. J.: Toxemia of pregnancy in the guinea pig. J. Exp. Med., *75*:539-547, 1942.

Ganaway, J. R., and Allen, A. M.: Obesity predisposes to pregnancy toxemia (ketosis) in guinea pigs. Lab. Anim. Sci., *21*:40-44, 1971.

Green, H. S. N.: Toxemia of pregnancy in the rabbit. I. Clinical manifestations and pathology. J. Exp. Med., *65*:809-832, 1937.

Green, H. S. N.: Toxemia of pregnancy in the rabbit. II. Etiologic considerations with especial reference to hereditary factors. J. Exp. Med., *67*:369-388, 1938.

Proliferative Ileitis (Wet Tail)

HOSTS

The recently weaned golden hamster, usually between the ages of 3 and 8 weeks, is the animal affected with proliferative ileitis (wet tail).

ETIOLOGY

The specific etiology is unknown, but the disease is possibly caused by several intestinal microorganisms, including one or more coliform serotypes and a *Vibrio* or *Campylobacter* spp. The diarrheal aspect may be caused by the coliforms and the epithelial hyperplasia by the *Vibrio* or *Campylobacter* spp.

TRANSMISSION

As the infectious agents are resident in the hamster gut, transmission is by the fecal-oral route.

PREDISPOSING FACTORS

Recently weaned hamsters (3 to 5 wk) are most often and most seriously affected. Improper diet, exposure to infected animals, and shipment stresses predispose to the development of acute enteritis and proliferative ileitis. Certain strains of hamsters may be more susceptible than others. Long-haired fawn and teddy bear hamsters appear to be highly susceptible.

CLINICAL SIGNS

Proliferative ileitis, an enteropathy occurring both endemically and epizootically, has both acute and chronic consequences. The acute condition involves increased irritability, diarrhea, dehydration, emaciation, and a high mortality. In less acute cases, survivors may die suddenly following ileal obstruction, intussusception, and impaction. Animals with intussusceptions may have a bloody diarrhea or prolapsed rectum.

NECROPSY SIGNS

Lesions of wet tail are most obvious in the short (18 mm) ileum and terminal jejunum. The ileal mucosa is thickened, causing the gut to be enlarged three to four times. The intestine is congested, often ulcerated, and contains a yellow fluid. Intussusceptions may occur during the acute or chronic phase. Perforation of the gut wall and a subsequent peritonitis may develop. Surviving hamsters may develop strictures, diverticula, multifocal hepatic abscesses, and adhesions.

DIAGNOSIS

Diagnosis of proliferative ileitis is based on the clinical observation of diarrhea and on the enlargement of the distal small intestine in weanling hamsters.

TREATMENT

Enteritis in colonies is prevented; it is rarely treated. The treatment described is for the pet or valuable animal when emergency, symptomatic treatment is indicated. Concerns in clinical enteritis are dehydration, acidosis, hypermotility, malnutrition, and septicemia secondary to intestinal necrosis.

As suggested treatments for diarrhea, lactated Ringer's solution is given SQ (5% to 15% body weight), a few ml kaolin with pectin (Kaopectate), or 1 drop/kg body weight each of the following: (1) dexamethasone (2 mg/ml); (2) neomycin (200 mg/ml)–methscopolamine (2 mg/ml) mixture, and (3) a liquid vitamin supplement.

PREVENTION

Proliferative ileitis is difficult to prevent in a group of affected, susceptible hamsters. High sanitary standards, absence of stressors, and cage filter covers may reduce the severity of an outbreak.

PUBLIC HEALTH SIGNIFICANCE

Proliferative ileitis in hamsters has no known public health significance.

BIBLIOGRAPHY

Boothe, A. D., and Cheville, N. F.: The pathology of proliferative ileitis of the golden Syrian hamster. Pathol. Vet. 4:31-44, 1967.
Jacoby, R. O., Obaldiston, G. W., and Jonas, A. M.:

Experimental transmission of atypical ileal hyperplasia of hamsters. Lab. Anim. Sci., *25*:465-473, 1975.
Jonas, A. M., Tomita, Y., and Wayand, D. S.: Enzootic intestinal adenocarcinoma in hamsters. J. Am. Vet. Med. Assoc., *147*:1102-1108, 1965.
Sheffield, F. W., and Beveridge, E.: Prophylaxis of "wet tail" in hamsters. Nature, *196*:294-295, 1962.
Wagner, J. E., Owens, D. R., and Troutt, H. F.: Proliferative ileitis of hamsters: Electron microscopy of bacteria in cells. Am. J. Vet. Res., *34*:249-252, 1973.

Salmonellosis

HOSTS

Salmonella spp affect a wide range of vertebrates. Guinea pigs are highly susceptible and may develop severe clinical disease. Mice are also quite susceptible and may carry latent, subclinical infections for long periods. Rabbits, hamsters, and rats are less often infected than guinea pigs and mice. Gerbils were long considered resistant to infection until a recent outbreak occurred in a research colony.

ETIOLOGY

Salmonella typhimurium and *Salmonella enteritidis* are the species most often isolated from research animal species.

TRANSMISSION

Transmission is fecal-oral through the ingestion of feces in contaminated greens and feed or by direct contact with the feces of carrier animals and humans. The organism can exist in the carrier state in the intestinal tract.

PREDISPOSING FACTORS

Among the several factors predisposing to the clinical disease are age, nutritional deficiencies, concomitant diseases, and nonspecific stress.

CLINICAL SIGNS

Salmonellosis in laboratory animals is usually an acute to subacute, fatal septicemia. Signs include anorexia, weight loss, light-colored, soft feces, conjunctivitis, dyspnea, and abortions. Chronic carriers and shedders exist and render elimination of the infection from a colony difficult.

NECROPSY SIGNS

Lesions of acute salmonellosis include enlargement, generalized congestion, and focal necrosis of the liver, spleen, lymphoid tissue, and intestine.

In subacute or chronic cases the spleen may be enlarged approximately two times, the liver and gut may be congested and enlarged, and yellow necrotic foci may be prominent in the viscera. Enlarged Peyer's patches may be seen through the serosal surface of the gut. A fibrinous peritonitis may occur.

DIAGNOSIS

Diagnosis of salmonellosis is based on the recovery of *Salmonella* spp from the spleen, blood, or feces. Recovery of *Salmonella* from the feces is often difficult unless 20 to 25 gm feces are cultured in an enteric broth such as selenite.

TREATMENT

Treatment of salmonellosis may suppress an epizootic to an enzootic infection, but elimination of carriers is difficult. Because of the major public health concern, colonies infected with *Salmonella* are eliminated, premises are sanitized, and clean animals are used for restocking.

PREVENTION

Rigid, high husbandry standards and the screening of new animals and animal care personnel will reduce the possibility of outbreaks. Birds, wild rodents, other vermin, and contaminated feed must be excluded from the colony.

PUBLIC HEALTH SIGNIFICANCE

Salmonellosis occurs in man and can be contracted from, or given to, laboratory animals. Animal care personnel should be periodically inspected for latent infections of *Salmonella*.

BIBLIOGRAPHY

Corazzola, S., Zanin, E., and Bersani, G.: Food poisoning in man following an outbreak of salmonellosis in rabbits. Vet. Ital., *22*:370-373, 1971.

Dutchie, R. C., and Mitchell, C. A.: *Salmonella enteritidis* infection in guinea pigs and rabbits. J. Am. Vet. Med. Assoc., *78*:27-41, 1931.

Ghosh, G. K., and Chatterjee, A.: Salmonellosis in guinea pigs, rabbits, and pigeons. Indian Vet. J., *37*:144-148, 1960.

Habermann, R. T., and Williams, F. P.: Salmonellosis in laboratory animals. J. Natl. Cancer Inst., *20*:933-941, 1958.

Innes, J. R. M., Wilson, C., and Ross, M. A.: Epizootic *Salmonella enteritidis* infection causing septic pulmonary phlebothrombosis in hamsters. J. Inf. Dis., *98*:133-141, 1956.

Margard, W. L., et al.: Salmonellosis in mice—diagnostic procedures. Lab. Anim. Care, *13*:144-165, 1963.

Olfert, E. D., Ward, G. E., and Stevenson, D.: *Salmonella typhimurium* infection in guinea pigs: Observations on monitoring and control. Lab. Anim. Sci., *26*:76-80, 1976.

Ray, J. P., and Mallick, B. B.: Public health significance of *Salmonella* infections in laboratory animals. Indian Vet. J., *47*:1033-1037, 1970.

Verma, N. S., and Sharma, S. P.: Salmonellosis in laboratory animals. Indian Vet. J., *46*:1101-1102, 1969.

Sendai Virus Infection

HOSTS

Mice, rats, hamsters, guinea pigs, and swine are hosts for Sendai virus infection. Mice and rats are most commonly and seriously affected.

ETIOLOGY

The Sendai virus is a RNA paramyxovirus-parainfluenza type 1. Many paramyxovirus-parainfluenza viruses share common antigens.

TRANSMISSION

Sendai virus causes a common and widespread enzootic and, less commonly, epizootic respiratory infection in rodent colonies. The mechanism of transmission is incompletely known, but passage may be by direct contact, contaminated fomites, or respiratory aerosol. Acutely infected weanling mice (4 to 6 wk) provide the principal reservoir for transmission in susceptible colonies. Latent, persistent infections may exist. Disease-free or specific pathogen-free (SPF) animals are highly susceptible.

PREDISPOSING FACTORS

Nonspecific stressors precipitate the clinical respiratory disease from the subclinical infection. Some inbred strains (129S and DBA) are susceptible to Sendai virus infection, but BALB/c, C57BL, and random bred strains are moderately resistant. Concurrent *Pasteurella pneumotropica* infections greatly increase morbidity and mortality.

CLINICAL SIGNS

Although subclinical enzootic disease may exist in a colony, stress or concurrent *P. pneumotropica* and *Mycoplasma pulmonis* infections can precipitate devastating outbreaks. Clinical signs of the acute infection include roughened hair coat, chattering, weight loss, dyspnea, decreased breeding efficiency, and variable mortality.

NECROPSY SIGNS

The acute pneumonitis is grossly evident as focal reddening, swelling, and firmness of the lungs. As the disease progresses to a subacute bronchopneumonia, focal consolidation occurs.

Histologically the acute phase is characterized by edema, hyperemia, and pneumonitis. In the subacute or resolution phase, focal areas of the bronchial and bronchiolar epithelium degenerate, while adenomatous hyperplasia and squamous metaplasia occur in the alveoli and distal bronchioles. Epithelial hyperplasia, metaplasia, and interstitial lymphocytic aggregations occur approximately 19 to 23 days after clinical onset.

DIAGNOSIS

The virus can be isolated from the saliva or lungs of hosts until antibodies develop. A hemagglutination inhibition (using human O or fowl erythrocytes) and a complement fixation test may be used to detect exposure to the Sendai virus.

PREVENTION

Prevention of Sendai virus infection, clinical or subclinical, involves selection of rodents from a Sendai-free source, the use of filter cage covers, and seromonitoring. The disease is highly contagious and difficult to control.

PUBLIC HEALTH SIGNIFICANCE

Little conclusive evidence exists that the Sendai virus isolated from rodents affects man. The antigens shared among the parainfluenza viruses confuse the diagnosis.

BIBLIOGRAPHY

Appell, L. H., et al.: Pathogenesis of Sendai virus infection in mice. Am. J. Vet. Res., *32*:1835-1841, 1971.

Fujiwara, K. F., et al.: Carrier state of antibody and viruses in a mouse breeding colony. Lab. Anim. Sci., *26*:153-159, 1976.

Makino, S., et al.: An epizootic of Sendai virus infection in a rat colony. Exp. Anim., *22*:275-280, 1972.

Parker, J. C., and Reynolds, R. K.: Natural history of Sendai virus infection in mice. Am. J. Epidemiol., *88*:112-125, 1968.

Parker, J. C., et al.: Enzootic Sendai virus infections in mouse breeder colonies within the United States. Science, *146*:936-938, 1964.

van der Veen, J., Poort, Y., and Birchfield, D. J.: Study of the possible persistence of Sendai virus in mice. Lab. Anim. Sci., *24*:48-50, 1974.

Ward, J. M.: Naturally occurring Sendai virus disease of mice. Lab. Anim. Sci., *24*:938-942, 1974.

Streptococcus pneumoniae Infection

HOSTS

Streptococcus pneumoniae affects a wide range of animals, but the guinea pig and rat are particularly susceptible.

ETIOLOGY

Streptococcus pneumoniae (the genera designations *Diplococcus* and *Pneumococcus* are also used with this organism) infections in the guinea pig are caused by several serotypes including Types III, IV, and XIX. Diplococcal types reported in rats include II, III, VIII, and XVI. Diplococci are Gram-positive, lancet-shaped cocci which often occur in pairs.

TRANSMISSION

Transmission of *S. pneumoniae* is by respiratory aerosol or direct contact. Clinically normal guinea pigs and rats may carry the organism in the upper respiratory passages. Contacts during shipment and subsequent stresses frequently precipitate the disease. Approximately 40% to 70% of a human population carries *Diplococcus* in the respiratory passages. Injections into the thoracic or abdominal cavities precipitate fulminating cases of fibrinopurulent peritonitis or pleuritis in carrier animals.

PREDISPOSING FACTORS

Losses are greater in the winter months, following shipment, and in animals on marginal diets. Carrier animals frequently succumb when stressed.

CLINICAL SIGNS

Carriers of *S. pneumoniae* exist. Signs of a respiratory infection include dyspnea, anorexia, weight loss, depression, coughing, sneezing, and death. Abortion of fetuses may occur. Rough hair coats and hematuria may be seen in rats.

NECROPSY SIGNS

Gross lesions of *S. pneumoniae* infection include seropurulent and fibrinopurulent pleuritis, pericarditis, peritonitis, meningitis, otitis media, and bronchopneumonia.

DIAGNOSIS

Diagnosis is established by observation of the Gram-positive, lancet-shaped cocci in pairs or short chains on a stained, direct smear of the inflammatory exudate. Histologically, unlike many bacterial diseases, large numbers of bacterial organisms can usually be seen in tissue sections. Recovery of *S. pneumoniae* on blood agar culture in the presence of 10% CO_2 confirms the diagnosis. The organism is alpha hemolytic and bile soluble and is inhibited by ethylhydrocupreine (Optochin), an antibiotic available in impregnated paper discs. *Streptococcus pneumoniae* ferments inulin. If the organism is mixed with a specific antiserum, a capsular swelling or quellung reaction occurs.

TREATMENT

Treatment in most cases is impractical because of the advanced stage of the clinical condition at detection. Oxytetracycline at 0.1 mg/ml in the drinking water for 7 days has controlled epizootics but not eliminated the carrier state. *Streptococcus pneumoniae* organisms are susceptible to ampicillin, bacitracin, chloromycetin, erythromycin, lincomycin, and methicillin, but extreme care should be taken in treating guinea pigs with antibiotics. Broad-spectrum antibiotics have less imbalancing effect on intestinal flora than do narrow spectrum antibiotics, particularly those that suppress Gram-positive bacteria.

PREVENTION

Good husbandry, elimination of carriers, and the reduction of environmental stressors reduce the possibility and severity of outbreaks.

PUBLIC HEALTH SIGNIFICANCE

Streptococcus pneumoniae may cause pneumonia, otitis media, sinusitis, and meningitis in man, but the development of the disease depends on the serotype involved and the host's resistance.

BIBLIOGRAPHY

Homburger, E., et al.: An epizootic of *Pneumococcus* type 19 infections in guinea pigs. Science, *102*:449-450, 1945.

Petrie, G. F.: Pneumococcal disease of the guinea pig. Vet. J., *89*:25-30, 1933.

Tucek, P. C.: Diplococcal pneumonia in the laboratory rat. Lab. Anim. Digest, 7:32-35, 1971.

Weisbroth, S. H., and Freimer, E. H.: Laboratory rats from commercial breeders as carriers of pathogenic pneumococci. Lab. Anim. Care, *19*:473-478, 1969.

Zydeck, F. A., Bennett, R. R., and Langham, R. F.: Subacute pericarditis in a guinea pig caused by *Diplococcus pneumoniae*. J. Am. Vet. Med. Assoc., *157*:1945-1947, 1970.

Streptococcus zooepidemicus Infection

HOSTS

Streptococcal organisms are found in a wide variety of hosts, but among laboratory species the guinea pig is most often clinically affected.

ETIOLOGY

A beta-hemolytic *Streptococcus* (Lancefield's Group C), *S. zooepidemicus*, is commonly involved in streptococcal disease in guinea pigs.

TRANSMISSION

Transmission of *S. zooepidemicus* can occur through cutaneous wounds (bites), via aerosol into the respiratory tract, or by the conjunctival or genital routes. The agent may also traverse the oral mucosa, possibly in the tonsil region, and invade cervical lymph nodes.

PREDISPOSING FACTORS

Poor husbandry and general stress factors predispose to streptococcal infections. Biting facilitates transmission, as possibly does the traumatic effect of overgrown teeth and dietary roughage.

CLINICAL SIGNS

Cervical lymphadenitis, a sporadic, pyogenic infection, involves abscessation and in some cases subsequent drainage of the cervical lymph nodes. This lymphadenopathy is known as "lumps."

Torticollis or "wry neck" results from an extension of a streptococcal otitis media to the inner ear. Respiratory involvement is manifested clinically by nasal discharge and signs of pneumonia, including dyspnea and cyanosis. A fatal septicemia with hematuria and hemoglobinuria may occur. Chronic infection can proceed to an acute septicemia if the host is stressed or an abscess ruptures spontaneously or is ruptured surgically. Animals with pneumonia and septicemia frequently abort.

NECROPSY SIGNS

Infection with *S. zooepidemicus* ranges from an acute, fatal septicemia to chronic, suppurative processes in the lymph nodes, thoracic and abdominal viscera, and middle and inner ears.

DIAGNOSIS

Diagnosis is based on the isolation of *Streptococcus* spp from affected tissues and blood. At 24 hours on blood agar, colonies are mucoid and 2 to 3 mm in diameter. They are surrounded by a zone of beta hemolysis and are easily confused with colonies of *Pasteurella multocida*. *Pasteurella multocida*, however, is Gram-negative.

TREATMENT

Concerns in therapy include the number of abscesses, presence of fistulous tracts, the potential for septicemia if the abscess is ruptured, and the antibiotic sensitivity of the organism. If surgical drainage is indicated, the animal is restrained with ketamine IM 25 mg/kg. The abscess is opened, drained, and flushed three times a day for 5 days with 2% chlorhexidine or tincture of iodine. With or without surgery, a systemic antibiotic, such as cephaloridine, is injected at 20 mg/kg IM per day for 5 to 7 days.

PREVENTION

Good husbandry practices, general preventive measures, and routine palpation for enlarged cervical lymph nodes reduce the possibility of streptococcal infection in guinea pigs. Affected guinea pigs are removed from the colony or treated with cephaloridine until complete resolution of abscesses occurs. In cases of epizootic *S. zooepidemicus* infections, with widespread pneumonia and septicemias, the entire colony should be eliminated.

PUBLIC HEALTH SIGNIFICANCE

Beta-hemolytic streptococci (Group A) have been recovered from human infections. *Streptococcus zooepidemicus*, however, is an animal pathogen.

BIBLIOGRAPHY

Corrado, A.: Streptococcal cervical lymphadenitis in guinea pigs. Nuova Vet., *39*:162-166, 1963.

Fraunfelter, F. C., et al.: Lancefield type C Streptococcal infections in Strain 2 guinea pigs. Lab. Anim., *5*:1-13, 1971.

Kohn, D. F.: Bacterial otitis media in the guinea pig. Lab. Anim. Sci., *24*:823-825, 1974.

Seastone, C. V.: Hemolytic streptococcus lymphadenitis in guinea pigs. J. Exp. Med., *70*:347-359, 1939.

Smith, W.: Cervical abscesses of guinea pigs. J. Path. Bacteriol., *53*:29-37, 1941.

Soave, O. A., and Diaz, J.: Cephaloridine treatment of cervical lymphadenitis in guinea pigs. Lab. Anim. Digest, *8*:60-62, 1973.

Steward, D. D., et al.: An epizootic of necrotic dermatitis in laboratory mice caused by Lancefield group C streptococci. Lab. Anim. Sci., *25*:296-302, 1975.

Wren, W. B., and Twiehaus, M. J.: Studies on the etiology, pathology, and control of cervical lymphadenitis in guinea pigs. Proc. Anim. Lab. Med. Abst., No. 157, 1968.

Tularemia

HOSTS

Tularemia has a wide host spectrum, and rodents and lagomorphs are highly susceptible and have been involved in epizootics. Tularemia is extremely rare in laboratory animals, although animals raised outdoors and exposed to ticks are much more likely to contract infections than are similar animals kept indoors in a home or laboratory.

ETIOLOGY

Francisella tularensis is a Gram-negative, pleomorphic, bipolar rod.

TRANSMISSION

Francisella tularensis is spread by direct contact through skin and via aerosol, ingestion, or blood-sucking arthropods, especially ticks.

PREDISPOSING FACTORS

The tularemia organism is highly infectious and affects otherwise healthy animals.

CLINICAL SIGNS

Tularemia is an acute, fatal septicemia. Clinical signs include depression, anorexia, ataxia, and death. The course in cottontails lasts approximately one week.

NECROPSY SIGNS

Signs of a septicemia include pulmonary congestion and consolidation, numerous pinpoint nacreous hepatic foci, and congestion of the liver and spleen. The white spots on the dark background of the congested liver and spleen resemble the Milky Way.

DIAGNOSIS

Diagnosis is based on necropsy findings, recovery of *F. tularensis* on cysteine-heart infusion media, or death of a guinea pig with lymphoid necrosis and septicemia 5 to 10 days after an intraperitoneal injection of the suspect's blood.

TREATMENT

Treatment is not indicated for animals. Tetracyclines and streptomycin are the antibiotics used to treat human infections.

PREVENTION

Exclusion of wild mammals and insect vectors from the colony are preventive measures. Bacterins have been developed. Hunters should avoid "lazy" cottontails.

PUBLIC HEALTH SIGNIFICANCE

Humans are very susceptible. As contamination of inapparent skin lesions results in human infections, persons handling suspect tissues should always wear gloves. Laboratory culture should only be attempted where adequate biohazard culture facilities are available. Cutaneous lesions, septicemias, and meningitis occur in cases of human tularemia. Bites of 2 and 3 host ticks are particularly dangerous, as ticks transmit the bacterium.

BIBLIOGRAPHY

Belding, D. L., and Merrill, B.: Tularemia in imported rabbits in Massachusetts. N. Engl. J. Med., *224*:1085-1087, 1941.

Eigelsbach, H. T., and Downs, C. H.: Prophylactic effectiveness of live and killed tularemia vaccines. J. Immunol., *87*:415-425, 1961.

Hoff, G. L., et al.: Tularemia in Florida: *Sylvilagus palustris* as a source of human infection. J. Wildl. Dis., *11*:560-561, 1975.

Moe, J. G., et al.: Pathogenesis of tularemia in immune and nonimmune rats. Am. J. Vet. Res., *36*:1505-1510, 1975.
Perman, V., and Bergeland, M. E.: A tularemia enzootic in a closed hamster breeding colony. Lab. Anim. Care, *17*:563-568, 1967.

Venereal Spirochetosis

HOSTS

Rabbits are susceptible to *Treponema cuniculi* infections.

ETIOLOGY

Treponema cuniculi is a slender, spiral-shaped bacterium.

TRANSMISSION

The spirochete is transmitted by direct contact, genital or extragenital. Exchange of bucks among breeders promotes dissemination of the organism.

PREDISPOSING FACTORS

Nonspecific stressors predispose to the clinical disease. Young rabbits, females, and rabbits in cold environments are predisposed to the clinical disease.

CLINICAL SIGNS

The clinical, epizootic disease is uncommon, although serologically positive sera are often encountered. Venereal spirochetosis is a cutaneous disease with raised, crusted, occasionally hemorrhagic, ulcerated foci on the external genitalia, perineal areas, and the face, particularly the nose. Affected rabbits remain alert, and the condition regresses after several weeks. If the prepuce is severely affected, apparent infertility may result.

NECROPSY SIGNS

The perineal and genital skin and the mucous membranes of the external genitalia are the tissues most often affected. Adjacent lymph nodes may be swollen.

DIAGNOSIS

The lesions of venereal spirochetosis in rabbits resemble those caused by trauma, dermatophytes, or ectoparasites. The use of the cardiolipin antigen Wassermann-type test, the Rapid Plasma Reagin (RPR) card test, will provide evidence of a *T. cuniculi* infection. The spirochetes in the epidermis, dermis, or lymph nodes may be demonstrated using dark field microscopy or a silver stain.

TREATMENT

Penicillin provides a cure. Penicillin is given intramuscularly at 40,000 units/kg per day for 3 to 5 days.

PREVENTION

Routine screening with the RPR card test provides an indication of exposure. Periodic examination of breeding does and bucks for cutaneous lesions will eliminate clinical carriers.

PUBLIC HEALTH SIGNIFICANCE

Although Rhesus monkeys and a chimpanzee have been experimentally infected, man is not believed to be susceptible to *T. cuniculi* infection.

BIBLIOGRAPHY

Chapman, M. P.: The use of penicillin in the treatment of spirochetosis (vent disease) of domestic rabbits. N. Am. Vet., *20*:740-742, 1947.
Clark, J. W., Jr.: Serological tests for syphilis in healthy rabbits. Br. J. Vener. Dis., *46*:191-197, 1970.
Small, J. D., and Newman, B.: Venereal spirochetosis of rabbits (rabbit syphilis) due to *Treponema cuniculi*: A clinical, serological, and pathological study. Lab. Anim. Sci., *22*:77-89, 1972.
Smith, J. L., and Pesetsky, B. R.: The current status of *Treponema cuniculi:* Review of the literature. Br. J. Vener. Dis., *43*:117-127, 1967.

Miscellaneous Conditions

ANOREXIA

Anorexia, or a reduction or cessation of food intake, may result from neophobia, water deprivation, climatic extremes, an unpalatable or improperly compounded diet, malocclusion, pain, toxemia, gastric hairball, territorial and behavioral traits, or mechanical factors that prevent access to feed. Clinical manifestations of anorexia are weight loss or failure to gain, agalactia with death or cannibalism of the litter, decreased resistance to disease, and death. Some problems involving anorexia can be overcome through the use of sweetened or preferred feeds (corn, fresh vegetables, sunflower seeds) or by mixing old and new feeds during a transition from one feed to another.

DYSTOCIA

Dystocias occur if the fetuses are too large for the pelvic canal or if the uterine musculature is weakened by toxins or infection. If the guinea pig is first bred after the pubic symphysis has ossified (around 9 mo), a subsequent dystocia is probable. Dystocia is associated with uterine hemorrhage, hemorrhage about the symphysis, exhaustion, toxemia, and death. Cesarean section or oxytocin (0.2 to 3.0 units/kg) may be used to alleviate a dystocia.

EPILEPTIFORM SEIZURES IN GERBILS

These hypnotic, cataleptic, or convulsive seizures in gerbils have a variable genetic threshold and are initiated by handling, environmental disturbances, or a novel environment. Approximately 20% of a gerbil population will exhibit the severe seizure pattern, which lasts from a few seconds to over a minute. The seizures are prevented either through the avoidance of novel environments or the use of diphenylhydantoin (Dilantin) or reserpine. The seizures apparently have no long-term adverse effect on the gerbil.

MALOCCLUSION

Malocclusion and tooth overgrowth result when, for genetic, dietary, infectious, or traumatic reasons, open-rooted teeth do not properly occlude and are therefore not eroded. Loss of an opposite tooth or prognathism commonly leads to tooth overgrowth.

In the rabbit the teeth commonly involved are the incisors, which, if not constantly worn down, will grow approximately 10 cm (4 in) per year. The cause of malocclusion in the rabbit is probably an autosomal recessive trait involving an abnormally short maxillary diastema. Overgrowth of the open-rooted cheek teeth of the rabbit is uncommon.

Malocclusion in guinea pigs, often difficult to detect, occurs in the open-rooted cheek teeth, usually the anterior cheek teeth. Other common laboratory rodents have multirooted molars that do not grow continuously.

Overgrown teeth result in trauma to the tongue and mouth and cause ptyalism (slobbers), anorexia, starvation, and death. The drooling around the mouth and onto the forequarters predisposes to moist dermatitis. Treatment of tooth overgrowth involves clipping (but not splitting or shattering) the teeth with a sharp instrument or dental burr. A split tooth will become impacted with food and eventually abscessed.

MASTITIS

Acute inflammation of the mammary gland is most often encountered during lactation when the milk provides an excellent medium for bacterial growth, the glands are pendulous, and the young traumatize the nipples. Mastitis is established when one or several genera of microorganisms (*Pasteurella, Klebsiella,* coliforms, *Streptococcus, Staphylococcus, Pseudomonas*) enter the gland via the blood stream, through a cutaneous lesion, or by the teat canal. Unsanitary conditions, biting young, and early weaning with mammary impaction predispose to mastitis. The affected gland becomes diffusely or focally enlarged, hyperemic, warm, and cyanotic. Depression and death from septicemia or toxemia follow. Young guinea pigs, however, are precocious and usually survive. Suckling young may also die from malnutrition or septicemia. Treatment of the acute condition includes hot pack application and systemic antibiotic therapy for 5 to 7 days.

MOIST DERMATITIS

Moist dermatitis, also known as "sore nose" in gerbils and "wet dewlap" and "hutch burn" in rabbits, is a superficial bacterial dermatitis caused by any one of a variety of organisms, including *Staphylococcus*, *Pseudomonas*, *Treponema*, *Streptococcus*, *Fusobacterium*, and *Corynebacterium*. Moist dermatitis is a frequent sequela to constant wetting, cutaneous abrasion, or puncture wounds. Conjunctivitis, drooling from malocclusion, use of water pans or crocks, chronic diarrhea, and damp cages predispose to moist dermatitis. Sore nose in gerbils may result from the irritation of constant burrowing activity. Prevention involves maintaining soft, clean bedding in a dry cage. Mice and other rodents may develop nasal dermatitis from eating through feeder openings too small or abrasive for their noses and mouths.

Moist dermatitis is treated by removing hair over and adjacent to the lesion, cleansing the area with an antiseptic soap, and administering topical or systemic antibiotics.

NITRATE TOXICITY

Nitrates enter the drinking water from fertilizer, organic matter, or manure runoff. Nitrates are present in many feed crops, including alfalfa, clover, orchard grass, timothy hay, and blue grass. Drought, high temperatures, cloudy days, chemical contaminants, and immaturity at harvest contribute to nitrate accumulation in plants. Rabbits may abort if exposed to low levels of nitrate ion in the drinking water. Nitrate analyses of feed and water should be obtained in suspect cases.

NONSPECIFIC ENTEROPATHY

Nonspecific enteropathies as defined here are enteric diseases caused by undetermined processes or enteric organisms which may or may not be normal events or residents of the intestine. The immediate cause of an enteropathy involves poorly understood alterations in the transport, secretory, or protective function of the intestinal mucosa. The alteration may involve an infectious or toxic agent, the overgrowth of a "normal" floral component, or an aberrant metabolic process.

Coccidiosis, salmonellosis, *Pseudomonas* infection, proliferative ileitis, and Tyzzer's disease are conditions of known etiology and are not discussed here; however, any of these diseases may occur simultaneously with a nonspecific enteropathy. Enteropathies of unknown etiology include mucoid enteropathy, acute cecitis, hemorrhagic cecitis, enterotoxemias, and certain protozoal infections. Mucoid enteropathy is discussed under a separate heading.

Neonatal and weanling diarrheas, which are common in rabbits, may arise when aberrant flora and fauna become established or proliferate in the relatively aseptic gut of the neonate. Also, prolonged administration of antibiotics affects the intestinal organisms and may precipitate an enteropathy. When an infectious agent is involved, transmission is fecal-oral.

Enteropathies are more often seen in stressed and weanling animals. The winter season, pregnancy, lactation, concurrent diseases, and miscellaneous stresses predispose adult animals to enteropathies. Clinical signs of an enteropathy vary from a transient, soft stool to chronic diarrhea to sudden death. Diarrhea, if it occurs, may be watery, soft, mucoid, or bloody. Affected animals exhibit weight loss, dehydration, malaise, teeth grinding, abdominal distention, and polydipsia. Gross necropsy signs vary from loose feces and congestion of the intestinal wall to hemorrhage, edema, and necrosis of various sections of the gut. Intestinal contents may appear normal, mucoid, watery, or dark. Mesenteric lymph nodes and the liver may be congested with pale foci of necrosis.

Diagnosis of an enteropathy is based on clinical and necropsy signs. A history involving poor husbandry, environmental stressors, and dietary manipulation or supplementation may establish causative conditions. Culture of the gut for bacteria or histologic examination of the affected mucosa for coccidia, other protozoa, and the Tyzzer's organism is indicated.

Enteropathies are prevented and rarely treated. The treatment described is symptomatic and intended for the pet or valuable research animal when emergency treatment is indicated. The concerns in treating clinical enteropathies include dehydration, gut hypermotility, enterotoxemia, antibiotic sensitivity, and malnutrition. As treatments, kaolin with pectin (Kaopectate) may be given orally or lactated Ringer's solution SQ (5% to 15% body weight). One drop/kg body weight of each of the following may be given, as indicated: dexamethasone (2 mg/ml); neomycin (200 mg/ml)–methscopolamine (2 mg/ml) mixture, and a vitamin supplement. This regimen is untested in laboratory animals, but may suggest a treatment.

High level husbandry standards and the reduction of predisposing factors, particularly dietary manipulation, are important in preventing enteropathies. Depending on the species involved, filter cage covers, self-cleaning wire cages, water bottles, and clean bedding, feed, cages, and animal handlers contribute to the elimination of nonspecific enteropathies. Dimetridazole (Emtryl) in the drinking water at 0.025% or 45 gm dimetridazole/50 gal drinking water administered during weaning (3 to 8 wk in rabbits) has been shown to reduce outbreaks of nonspecific enteropathies in rabbits. The mechanism of action is unknown, although it may involve inhibition of protozoal growth. Dimetridazole is approved only for use in turkeys. Such antimicrobial supplementations are temporary substitutes for good husbandry. In several trials, high fiber diets (16% to 25%) have reduced the incidence of enteropathies in rabbit colonies.

NUTRITIONAL IMBALANCES

Provision of an adequate diet should underlie all efforts to maintain the health of laboratory animals. Shortcut, discount, outdated, or improperly formulated diets should always be suspect. The nutritional requirements of many research animal species have been thoroughly investigated and extensively recorded. These requirements are available from feed companies or from the National Research Council publications listed in the bibliography to Chapter 1.

With the exception of ascorbic acid deficiency in guinea pigs and net energy, salt, water, and protein deficiencies in all species, malnutrition is uncommon in laboratory animals. Nutritional problems uncommonly encountered include vitamin D, calcium, and phosphorus imbalances in rabbits (atherosclerosis), guinea pigs (metastatic calcification), and young rodents (rickets); vitamin A (hydrocephalus, prenatal death) and vitamin E (muscular dystrophy, prenatal mortality, seminiferous tubule degeneration) deficiencies in rabbits and rodents; and certain mineral or amino acid deficiences in several species.

Subclinical nutritional deficiencies, excesses, or imbalances may be more common than suspected, as the signs are often obscured by secondary bacterial infections or metabolic disorders. The importance of nutritional imbalances lies more in the predisposing role than in the causation of primary deficiencies or toxic lesions.

When nutritional studies are not involved, laboratory animal species should be fed a complete, fresh, wholesome, palatable, clean, pelleted diet. Supplementation with straw, greens, salt blocks, and antibiotics should be undertaken only with an understanding of possible benefits and adverse consequences. Also, the diet that might be preferred by the animal is not necessarily the proper diet for that species.

Primary nutritional imbalances may be manifested as weight loss or failure to gain, increased susceptibility to disease, hair loss, prenatal mortality, agalactia, infertility, anemia, deformed bones, central nervous system abnormalities, or a reluctance to move. Individual pet animals, more frequently than laboratory animals, experience nutritional deficiencies because fresh, complete diets are less available to the pet owner than to the large research laboratory.

RINGTAIL

If unweaned rats 7 to 15 days old are housed in low (20% or less) ambient humidity, they may develop one or more annular constrictions of the tail. These constrictions progress to edema, inflammation, and necrosis of the tail distal to the constriction. If the tail is sloughed, the stump usually heals without complications.

Ringtail is associated with low humidity and a related, aberrant response of the temperature-regulating vessels in the tail of the neonatal, poikilothermic rat. Conditions that cause lowered ambient humidity (wire cages, hygroscopic bedding, excessive ventilation) predispose to ringtail. Ringtail is prevented by providing a solid bottom nesting box or cage with both adequate bedding and nesting material. The relative humidity of the room should remain at approximately 50%. Ringtail is most often seen between the months of November and May.

SPLAY LEG

Splay leg is a descriptive term applied to a variety of inherited or acquired abnormalities in which the rabbit is unable to adduct one or more limbs. These abnormalities include improper development of the spine, pelvis, coxofemoral junction, or long bones. The muscles of the affected limbs may remain functional or become partially or totally paralyzed. Unlike paralysis associated with vertebral fracture or luxation, splay leg usually has a familial inheritance pattern. Splay leg may be first evident at a few days or after several months of age.

TRICHOBEZOARS (GASTRIC HAIRBALLS)

Gastric hairballs occur commonly in rabbits and rarely in other laboratory species. As a result of self-grooming or barbering, a mass of hair accumulates in the stomach. If the mass becomes large enough, pyloric obstruction occurs, and weight loss, agalactia, depression, and death follow.

Palpation for a suspected hairball may result in a ruptured liver, especially if the liver has undergone fatty change. Radiographic diagnosis is possible if a barium suspension or air is passed by stomach tube. The barium or air passes between the mass and gastric wall and outlines the hairball.

The specific cause of excessive grooming and hair swallowing has not been determined, but boredom or deficiencies of copper, fiber, or magnesium have been suggested. Alfalfa hay, mineral oil (10 ml), a liquid detergent, or a gastrotomy might be attempted to remove the hairball.

ULCERATIVE PODODERMATITIS

Ulcerative pododermatitis in rabbits, commonly known as "sore hocks," occurs unilaterally or bilaterally on the plantar metatarsal or metacarpal surfaces. This common clinical disease consists of focal, scab-covered, often hemorrhagic, cutaneous ulcers.

Factors predisposing to sore hocks include reduced plantar fur pad thickness (inherited or through wetting), thumping and bruising, pressure or decubital ulceration, lack of movement in a small cage, or abrasions from irregular cage flooring. Secondary bacterial infection may complicate treatment.

Sore hocks, which can occur in any type of cage or confinement device, is prevented by housing rabbits in clean, dry cages on soft, clean, dry bedding. Treatment includes cleansing with an antiseptic soap and the application of a topical antiseptic such as tincture of iodine.

Ulcerative pododermatitis in guinea pigs (bumblefoot) is a chronic dermatitis that progresses to arthritis. The swelling, which usually involves the forefeet, may be 2 to 3 cm in diameter, hairless, and scabbed. The inflammatory process infiltrates joints, tendon sheaths, and connective tissue but seldom produces a pus pocket that can be opened and drained. The condition is most common in heavy animals held for long periods on an abrasive flooring. *Staphylococcus aureus*, the usual causative agent, probably enters the foot through a cutaneous wound. Housing animals in clean, soft, or smooth-floored cages reduces the incidence of ulcerative pododermatitis. Systemic amyloidosis in the guinea pig is associated with chronic staphylococcal infection. The condition rarely responds to treatment with antibiotics.

VERTEBRAL LUXATION OR FRACTURE

Fracture or luxation of the lumbar spine is most common in the rabbit, which, if dropped, allowed to thrash during attempted restraint, or excited in the cage, may develop a fracture or luxation of the lower lumbar spine. The degree and duration of the resulting paresis or paralysis and loss of bladder and sphincter control depend on the severity and location of the cord lesion. If radiography or palpation reveals displacement of the vertebral canal and probable cord laceration, the prognosis is poor, and recovery is unlikely. If the cord is only edematous or locally inflamed, recovery may follow cage rest.

MISCELLANEOUS RABBIT DISEASES

Alpen, G. R., and Maerz, K.: The incidence of a pathogenic strain of *Pseudomonas* in a rabbit colony. J. Inst. Anim. Tech., *20*:72-74, 1969.

Arendar, G. M., and Milch, R. A.: Splay leg—a recessively inherited form of femoral neck anterversion, femoral shaft torsion and subluxation of the hip in the laboratory lop rabbit. Clin. Ortho., *44*:221-229, 1966.

Beamer, R. H., and Penner, L. R.: Observations on the life history of a rabbit cuterebrid, the larvae of which may penetrate the human skin. J. Parasitol., *28*:25, 1942.

Beattie, J. M., Gates, A. G., and Donaldson, M. A.: An epidemic disease in rabbits resembling that produced by *B. necrosis* (Schmorl) but caused by an aerobic bacillus. J. Pathol. Bacteriol., 18:34-36, 1913.

Bryner, J. H., et al.: Infectivity of three *Vibrio fetus* biotypes for gallbladder and intestines of cattle, sheep, rabbits, guinea pigs, and mice. Am. J. Vet. Res., *32*:465-470, 1971.

Christensen, L. R., Bond, E., and Matanic, B.: Pockless rabbit pox. Lab. Anim. Care, *17*:281-296, 1967.

Cutlip, R. C., et al.: Susceptibility of rabbits to infection with Lancefield's group E streptococci. Cornell Vet., *61*:607-616, 1971.

Dade, A. W., et al.: An epizootic of cerebral nematodiasis in rabbits due to *Ascaris columnaris*. Lab. Anim. Sci., *25*:65-69, 1975.

Daniels, J. J. H. M.: Enteral infections with *Pasteurella pseudotuberculosis*. Br. Med. J., *2*:997, 1961.

Dixon, C. F.: Infection of domestic rabbits and gerbils with *Trichostrongylus affinis*. J. Parasitol., *51*:299, 1965.

Duncan, C. L., and Strong, D. H.: Experimental production of diarrhea in rabbits with *Clostridium perfringens*. Can. J. Microbiol., *15*:765-816, 1969.

Gray, M. L., Singh, C., and Thorp, F.: Abortions, stillbirths, early death of young in rabbits by *Listeria monocytogenes*. II. Oral exposure. Proc. Soc. Exp. Biol. Med., *89*:169-175, 1955.

Green, H. S. N.: Rabbit pox. I. Clinical manifestations and course of the disease. J. Exp. Med., *60*:427-440, 1934.

Hagen, K. W.: Disseminated staphylococcic infection in young domestic rabbits. J. Am. Vet. Med. Assoc., *142*:1421-1422, 1963.

Harcourt, R. A.: Toxoplasmosis in rabbits. Vet. Rec., *81*:191-192, 1967.

Harkins, M. J., and Saleeby, E.: Spontaneous tuberculosis of rabbits. J. Infec. Dis., *43*:554-556, 1929.

Haust, M. D., and Greer, J. C.: Mechanism of calcification in spontaneous aortic arteriosclerotic lesions of the rabbit. Am. J. Pathol., *60*:329-346, 1970.

Holmes, R. G.: Listeriosis in rabbits. Vet. Rec., *73*:791, 1961.

Kaufmann, A. F., et al.: Pseudopregnancy in the New Zealand white rabbit: necropsy findings. Lab. Anim. Sci., *21*:865-869, 1971.

Kruckenberg, S. M.: Nitrate induced abortions in rabbits: observations of field and laboratory cases. Abst. #23, 25th Annual Session, AALAS, Cincinnati, 1974.

Lopushinsky, T.: Myiasis of nesting cottontail rabbit. J. Wildl. Dis., *6*:98-100, 1970.

McDonald, R. A., and Pinheiro, A. F.: Water chlorination controls *Pseudomonas aeruginosa* in a rabbitry. J. Am. Vet. Med. Assoc., *151*:863-864, 1967.

Mack, R.: Disorders of the digestive tract of domesticated rabbits. Vet. Bull., *32*:1-8, 1965.

Millen, J. W., and Dickson, A. D.: The effect of vitamin A upon the cerebrospinal fluid pressures of young rabbits suffering from hydrocephalus due to maternal hypovitaminosis. Br. J. Nutr., *11*:440-446, 1957.

Moon, H. W., et al.: Intraepithelial *Vibrio* associated with acute typhlitis of young rabbits. Vet. Pathol., *11*:313-329, 1974.

Nettles, V. F., et al.: An epizootic of cerebrospinal nematodiasis in cottontail rabbits. J. Am. Vet. Med. Assoc., *167*:600-604, 1975.

Oloufa, M. M., Bogart, R., and McKenzie, F. F.: Effect of environmental temperatures and the thyroid gland on fertility in the male rabbit. Fertil. Steril., *2*:223-228, 1951.

Patton, N. M.: Cutaneous and pulmonary aspergillosis in rabbits. Lab. Anim. Sci., *25*:347-350, 1975.

Pollock, S.: Slobbers in the rabbit. J. Am. Vet. Med. Assoc., *119*:443-444, 1951.

Renquist, D., and Soave, O.: Staphylococcal pneumonia in a laboratory rabbit: an epidemiologic follow-up study. J. Am. Vet. Med. Assoc., *155*:1221-1223, 1969.

Ringler, D. H., and Abrams, G. D.: Nutritional muscular dystrophy and neonatal mortality in a rabbit breeding colony. J. Am. Vet. Med. Assoc., *157*:1928-1934, 1970.

Roe, F. J. C., and Stiff, A. L.: Fracture dislocation of lumbar spine occurring spontaneously in rabbits. J. Amin. Tech. Assoc., *12*:92-94, 1962.

Rollins, W. C., and Casady, R. B.: An analysis of preweaning deaths in rabbits with special emphasis on enteritis and pneumonia. Anim. Prod., *9*:87-92, 1967.

Savage, M. L., et al.: An epizootic of diarrhea in a rabbit colony. Pathology and bacteriology. Can. J. Comp. Med., *37*:313-319, 1973.

Schenk, E. A., Gaman, E., and Feigenbaum, A. S.: Spontaneous aortic lesions in rabbits. I. Morphologic characteristics. Circ. Res., *19*:80-88, 1966.

Shotts, E. B., Jr., et al.: Leptospirosis in cottontail and swamp rabbits of the Mississippi Delta. J. Wildl. Dis., *7*:115-118, 1971.

Smith, H. W.: Observations on the flora of the alimentary tract of animals and factors affecting its composition. J. Pathol. Bacteriol., *89*:95-122, 1965.

Smith, H. W.: The antimicrobial activity of the stomach contents of suckling rabbits. J. Pathol. Bacteriol., *91*:1-9, 1966.

Smith, H. W.: The development of the flora of the alimentary tract in young animals. J. Pathol. Bacteriol., *90*:495-513, 1965.

Snyder, S. B., et al.: Disseminated staphylococcal disease in laboratory rabbits (*Oryctolagus cuniculus*). Lab. Anim. Sci., *26*:86-88, 1976.

Sollod, A. E., Hayes, T. J., and Soulsby, E. J. L.: Parasitic development of *Obeliscoides cuniculi* in rabbits. J. Parasitol., *54*:129-132, 1968.

Sommerville, R. J.: Distribution of some parasitic nematodes in the alimentary tract of sheep, cattle, and rabbits. J. Parasitol., *49*:593-599, 1963.

Stevenson, R. G., Palmer, N. C., and Finley, G. G.: Hypervitaminosis D in rabbits. Can. Vet. J., *17*:54-57, 1976.

Taylor, J., Williams, M. P., and Payne, J.: Relation of rabbit gut reaction to enteropathogenic *Escherichia coli*. Br. J. Exp. Pathol., *42*:43-52, 1961.

Templeton, G. S.: Treatment for paralyzed hindquarters. Am. Rabbit J., *16*:155, 1946.

Vetesi, F., and Kemeres, F.: Studies on listeriosis in pregnant rabbits. Acta Vet. Acad. Sci. Hung., *17*:27-38, 1967.

Wagner, J. L., Hackel, D. B., and Samsell, A. G.: Spontaneous deaths in rabbits resulting from gastric trichobezoars. Lab. Anim. Sci., *24*:826-830, 1974.
Weber, H. W., et al.: Cardiomyopathy in crowded rabbits. A preliminary report. S. Afr. Med. J., *47*:1591-1595, 1973.
Weisbroth, S. H., and Scher, S.: Naturally occurring hypertrophic pyloric stenosis in the domestic rabbit. Lab. Anim. Sci., *25*:355-360, 1975.
Weisbroth, S. H., and Ehrman, L.: Malocclusion of the rabbit. J. Hered., *58*:245-246, 1967.
Whitney, J. C.: Treatment of enteric disease in the rabbit. Vet. Rec., *95*:533, 1974.
Yuill, T. M., and Hanson, R. P.: Coliform enteritis of cottontail rabbits. J. Bacteriol., *89*:1-8, 1965.
Zeman, W. V., and Fielder, F. G.: Dental malocclusion and overgrowth in rabbits. J. Am. Vet. Med. Assoc., *155*:1115-1119, 1969.

MISCELLANEOUS GUINEA PIG DISEASES

Aldred, P., et al.: The isolation of *Streptobacillus moniliformis* from the cervical abscesses of guinea pigs. Lab. Anim., *8*:275-277, 1974.
Bishop, L. M.: Study of an outbreak of pseudotuberculosis in guinea pigs (cavies) due to *B. pseudotuberculosis rodentium*. Cornell Vet., *22*:1-9, 1932.
Cotchin, E.: A natural case of tuberculosis in a guinea pig. Vet. Rec., *56*:437, 1944.
Edwards, J. J.: Prenatal loss of fetuses and abortion in guinea pigs. Nature, *210*:223-224, 1960.
Gupta, B. N., Conner, G. H., and Meyer, D. B.: Osteoarthritis in guinea pigs. Lab. Anim. Sci., *22*:362-368, 1972.
Gupta, B. N., Langham, R. F., and Conner, G. H.: Mastitis in guinea pigs. Am. J. Vet. Res., *31*:1703-1707, 1970.
Juhr, N. C., and Obi, S.: Uterine infection in guinea pigs. Z. Versuchstierkd., *12*:383-387, 1970.
Maynard, L. A., et al.: Dietary mineral interrelationships as a cause of soft tissue calcification in guinea pigs. J. Nutr., *64*:85-97, 1958.
Plank, J. S., and Irwin, R.: Infertility of guinea pigs on sawdust bedding. Lab. Anim. Care, *16*:9-11, 1966.
Porter, D. A., and Otto, G. F.: The guinea pig nematode, *Paraspidodera uncinata*. J. Parasitol., *20*:323, 1934.
Sparschu, G. L., and Christie, R. J.: Metastatic calcification in a guinea pig colony: a pathologic survey. Lab. Anim. Care, *18*:520-526, 1968.
Taylor, J. L., et al.: Chronic pododermatitis in guinea pigs, a case report. Lab. Anim. Sci., *21*:944-945, 1971.
Townsend, G. H.: The guinea pig: general husbandry and nutrition. Vet. Rec., *96*:451-454, 1975.
Wright, J. E., and Seibold, J. R.: Estrogen contamination of pelleted feed for laboratory animals: effect on guinea pig reproduction. J. Am. Vet. Med. Assoc., *132*:258-261, 1958.

MISCELLANEOUS HAMSTER DISEASES

Frisk, C. S., Wagner, J. E., and Owens, D. R.: Streptococcal mastitis in golden hamsters. Lab. Anim. Sci., *26*:97-98, 1976.
Homburger, F., Editor: Pathology of the Syrian Hamster. Vol. 16. Progress in Experimental Tumor Research. New York, S. Karger, 1972.
Nelson, W. B.: Fatal hairball in a long haired hamster. Vet. Med. Small Anim. Clin., *70*:1193, 1975.
Small, J. D.: Fatal enterocolitis in hamsters given lincomycin hydrochloride, Lab. Anim. Care, *18*:411-420, 1968.
Verster, A.: The golden hamster as a definitive host of *Taenia solium* and *Taenia saginata*. Onderstepoort J. Vet. Res., *41*:23-28, 1974.

MISCELLANEOUS GERBIL DISEASES

Kaplan, H., et al.: Development of seizures in the Mongolian gerbil (*Meriones unguiculatus*). J. Comp. Physiol. Psychol., *81*:267-273, 1972.
Loew, F. M.: A case of overgrown mandibular incisors in a Mongolian gerbil. Lab. Anim. Care, *17*:137-139, 1967.
Loskota, W. J., et al.: The gerbil as a model for the study of the epilepsies. Seizure patterns and ontogenesis. Epilepsia, *15*:109-119, 1974.
Norris, M. L., and Adams, C. E.: Incidence of cystic ovaries and reproductive performance in the Mongolian gerbil, *Meriones unguiculatus*. Lab. Anim., *6*:337-342, 1972.
Peckham, J. C., et al.: Staphylococcal dermatitis in Mongolian gerbils (*Meriones unguiculatus*). Lab. Anim. Sci., *24*:43-47, 1974.
Thiessen, D. D., Lindzey, G., Friend, H. C.: Spontaneous seizures in the Mongolian gerbil, *Meriones unguiculatus*. Psychom. Sci., *11*:227-228, 1968.
Wexler, B. C., et al.: Spontaneous arteriosclerosis in male and female gerbils (*Meriones unguiculatus*). Atherosclerosis, *14*:107-119, 1971.

MISCELLANEOUS MOUSE DISEASES

Ball, C. R., and Williams, W. L.: Spontaneous and dietary-induced cardiovascular lesions in DBA mice. Anat. Rec., *152*:199-209, 1965.
Beck, R. W.: The control of *Pseudomonas aeruginosa* in a mouse breeding colony by the use of chlorine in the drinking water. Lab. Anim. Care, *13*:41-45, 1963.
Brennan, P. C., et al.: *Citrobacter freundii* associated with diarrhea in laboratory mice. Lab. Anim. Care, *15*:266-275, 1965.
Cook, I.: Reovirus type 3 infection in laboratory mice. Aust. J. Exp. Biol. Med. Sci., *41*:651-659, 1963.
Deringer, M. K., Dunn, T. B., and Heston, W. E.: Results of exposure of strain C3H mice to

chloroform. Proc. Soc. Exp. Biol., *83*:474-479, 1953.

Freundt, E. A.: Arthritis caused by *Streptobacillus moniliformis* and pleuropneumonia-like organisms in small rodents. Lab. Invest., *9*:1358-1375, 1959.

Galloway, J. H.: Antibiotic toxicity in white mice. Lab. Anim. Care, *18*:421-425, 1968.

Habermann, R. T., and Williams, F. P., Jr.: Treatment of female mice and their litters with piperazine adipate in the drinking water. Lab. Anim. Care, *13*:41-45, 1963.

Harkness, J. E., and Wagner, J. E.: Self mutilation in mice associated with otitis media. Lab. Anim. Sci., *25*:315-318, 1975.

Hoag, W. G.: Oxyuriasis in laboratory mouse colonies. Am. J. Vet. Res., *22*:150-153, 1961.

Litterst, C. L.: Mechanically self-induced muzzle alopecia in mice. Lab. Anim. Sci., *24*:806-809, 1974.

Loosli, J. K.: Primary signs of nutritional deficiencies of laboratory animals. J. Am. Vet. Med. Assoc., *142*:1001-1004, 1963.

Lussier, G., and Loew, F. M.: An outbreak of hexamitiasis in laboratory mice. Can. J. Comp. Med., *34*:350-353, 1970.

Riley, V.: *Eperythrozoon coccoides.* Science, *146*:921-923, 1964.

Stowe, H. D., Wagner, J. L., and Pick, J. R.: A debilitating fatal murine dermatitis. Lab. Anim. Sci., *21*:892-897, 1971.

Wagner, J. E., and Johnson, D. R.: Toxicity of Dichlorvos for laboratory mice LD_{50} and effect on serum cholinesterase. Lab. Anim. Care, *20*:45-47, 1970.

Wensinck, F., Ven Bekku, D. W., and Renaud, H.: The prevention of *Pseudomonas aeruginosa* infection in irradiated mice and rats. Radiat. Res., *7*:491-499, 1957.

MISCELLANEOUS RAT DISEASES

Ash, G. W.: An epidemic of chronic skin ulceration in rats. Lab. Anim., *5*:115-122, 1971.

Beare-Rogers, J. L., and McGowan, J. E.: Alopecia in rats housed in groups. Lab. Anim., *7*:237-238, 1973.

Berg, B. N., and Simms, H. S.: Nutrition and longevity in the rat. II. Longevity and onset of disease with different levels of food intake. J. Nutrition, *71*:255-263, 1960.

Blackmore, D. K., and Francis, R. A.: The apparent transmission of *Staphylococcus* of human origin to laboratory animals. J. Comp. Pathol., *80*:645-651, 1970.

Cotchin, E., and Roe, F. J. C., Editors: Pathology of Laboratory Rats and Mice. Philadelphia, F. A. Davis, 1967.

Dikshit, P. K., and Sriramachari, S.: Caudal necrosis in suckling rats. Nature, *181*:63-64, 1958.

Ford, A. C., Jr., and Murray, T. J.: Studies on *Haemobartonella* infection in the rat. Can. J. Microbiol., *5*:345-350, 1959.

Foster, H. L.: Comparison of epizootic diarrhea of suckling rats and a similar condition in mice. J. Am. Vet. Med. Assoc., *133*:198-201, 1958.

Geil, R. G., Davis, C. L., and Thompson, S. W.: Spontaneous ileitis in rats—a report of 64 cases. Am. J. Vet. Res., *22*:932-936, 1961.

Haley, A. J.: Host specificity of the rat nematode, *Nippostrongylus muris.* Am. J. Hyg., *67*:331-349, 1958.

Harr, J. R., Tinsley, I. J., and Weswig, P. H.: Haemophilus isolated from a rat respiratory epizootic. J. Am. Vet. Med. Assoc., *155*:1126-1130, 1969.

Hottendorf, G. H., Hirth, R. S., and Peer, R. L.: Megaloileitis in rats. J. Am. Vet. Med. Assoc., *155*:1131-1135, 1969.

Jacoby, R. O., Bhatt, P. N., and Jonas, A. M.: Pathogenesis of sialodacryoadenitis in gnotobiotic rats. Vet. Pathol., *12*:196-209, 1975.

Jones, L. P.: Purulent panophthalmitis in laboratory rats. J. Am. Vet. Med. Assoc., *135*:502-503, 1959.

Lynch, J. J., and Katcher, A. H.: Human handling and sudden death in laboratory rats. J. Nerv. Ment. Dis., *159*:362-365, 1974.

Magee, P. N.: Pathological changes in old rats in relation to chronic toxicity tests. Lab. Animals Centre, M.R.C. Lab., Collected Papers, *8*:59-68, 1959.

Skold, B. H.: Chronic arthritis in the laboratory rat. J. Am. Vet. Med. Assoc., *138*:204-207, 1961.

Timmons, E. H.: Dichlorvos effects on estrous cycle onset in the rat. Lab. Anim. Sci., *25*:45-47, 1975.

Totton, M.: Ringtail in new born Norway rats. A study of the effect of environmental temperature and humidity on incidence. J. Hyg., *56*:190-196, 1958.

Chapter 6

Case Reports

These 40 case reports were taken, with minor modifications, from the case report files of the Research Animal Diagnostic Laboratory, University of Missouri-Columbia. These cases reveal the complex patterns of disease in laboratory animals and demonstrate how field cases differ considerably from textbook descriptions. Short answers are supplied at the end of the section.

Rabbits

CASE 1

A student's science project involved restraining rabbits in a long, narrow, solid plastic box. Throughout the day the rabbits were wetted with urine. Following one week of this confinement, 2 of 5 rabbits developed cutaneous ulcers on the plantar surfaces of the hocks.

a. What recommendations would you make to the student to prevent the recurrence of this problem?
b. How would you treat the affected animals?
c. Why might dampness, especially dampness due to urine and feces, predispose to sore hocks?

CASE 2

A pet rabbit was brought to a veterinary clinic for examination. The rabbit had a pendulous abdomen, diarrhea, and cachexia.

a. What diagnostic procedures would contribute to a diagnosis?
b. Necropsy examination of the rabbit revealed an enlarged liver with numerous yellowish spots. How could you differentiate hepatic coccidiosis from *Taenia* (*Cysticercus*) *pisiformis* infection and tularemia?
c. What is the causative organism of hepatic coccidiosis?

CASE 3

A rabbit producer had two continuing problems in her rabbitry. The more serious problem involved a recurring, epizootic, highly fatal diarrhea in young rabbits. The rabbits were raised in wooden framed, wire cages off the ground. Crocks, cans, and pans were used for feeding and watering. The diarrhea was watery and green-brown. Adults were not affected. The second problem was a herd infertility occurring during the hot, humid months of summer.

a. What diagnostic procedures would you utilize to determine the cause of the enteritis?
b. What recommendations would you make to prevent the recurrence of the enteritis?
c. What is a possible cause of the herd infertility?

CASE 4

A backyard rabbitry husbandryman brought a 2-month-old buck with marked torticollis (wry neck) and nystagmus to the clinic. Although the animal appeared in good flesh and managed to eat in spite of the severe torticollis, the owner was worried about the communicability of the disease to other rabbits. Postmortem gross examination of the tympanic bullae, inner ears, and brain revealed no lesions. No microorganisms were cultured from the inner ear, meninges, or cerebrospinal fluid.

a. Is chronic otitis media or interna frequently encountered in young rabbits?
b. What are the diagnostic alternatives to otitis interna as a cause of wry neck in rabbits?
c. How would you proceed toward a definitive diagnosis of this case?

CASE 5

A young lady, a good client with a kennel of show dogs, consults with you regarding her four rabbits, which recently developed a mucopurulent nasal discharge and crusts in the ear canals.

a. What organism would you expect to isolate from the nasal passages?
b. Suggest a treatment for rhinitis in rabbits.
c. What agent would you probably find in a scraping from the ear canal? What control methods would you recommend for the ear problem?

CASE 6

An obese, 3-year-old, pregnant (27 days) doe with a history of sudden death was submitted for necropsy examination. The rabbit had been fed a diet of table scraps and milk. Gross lesions included multifocal, small (1 to 2 mm) pits on the cortical surface of the kidneys, 10 dead fetuses, and a light tan liver.

a. What agent is the probable cause of the renal lesions? How is this agent transmitted?
b. What is the probable cause of death? What husbandry steps are required to prevent the occurrence of this metabolic disease in a rabbitry?
c. How long is the rabbit's gestation period?

CASE 7

A dead adult doe had blue-green discoloration and matting of the fur on the entire ventrum. The animal had been frequently exposed to water and urine in the cage. Histologic examination of the skin revealed a severe, acute dermatitis characterized by multiple foci of epidermal ulceration and a diffuse inflammatory cell infiltration of the subcutis.

a. The blue-green discoloration of the fur suggests which bacterial organism may be present on the rabbit?
b. How can this bacterium be eliminated from a rabbitry?
c. What experimental manipulation may precipitate a *Pseudomonas* septicemia in mice?

CASE 8

An adult doe with a history of chronic weight loss was presented to the clinic. The rabbit had lost two successive litters one week postpartum. Radiographs failed to reveal internal abnormalities.

a. What are causes of cachexia in rabbits?
b. Would a gastric hairball have a role in the death of the litters?
c. Is there any treatment for a gastric hairball?

CASE 9

An adult, 4-year-old doe had a sanguineous, vulvar discharge. Catheterization of the urethra revealed a normal urine.

a. What organ was most probably affected?
b. How would you confirm your diagnosis?
c. What are the clinical signs of uterine neoplasia in rabbits?

CASE 10

An adult, female New Zealand White rabbit was submitted for clinical examination. The doe had delivered 10 young 4 weeks previously. All were healthy. The rabbit had focal areas of dry skin and alopecia. Under the skin in the inguinal region were four subcutaneous nodules 0.5 to 2 cm in diameter. On incision one nodule contained a caseous, inspissated material. *Staphylococcus aureus* was isolated from the abscess.

a. How would you treat the abscesses?
b. How would you proceed toward a definitive diagnosis of the cutaneous alopecia?
c. How is acariasis treated in rabbits?

CASE 11

An adult male rabbit was submitted for necropsy examination. The animal had a purulent conjunctivitis and rhinitis, and wet dewlap.

a. What are the causes of wet dewlap? How is this condition prevented and treated?
b. How would a *Pasteurella multocida* organism reach the orbit from the nasal passage?
c. How would you treat the conjunctivitis?

CASE 12

An adult doe was submitted for clinical examination. The rabbit had an irregularly shaped, red-purple, soft mass (4 × 5 cm) on the left forefoot. The rabbit was cyanotic and depressed.

a. You are the practitioner. How would you proceed to diagnose this case?
b. The mass on the foot was first diagnosed as a myxoma and then as a fibroma. What is the probable etiology of the neoplasm?
c. The leg was amputated. What is the probability of regression in this neoplastic condition?

CASE 13

A 2-month-old doe was submitted with a feces-soiled perineum. The attending clinician noted that the animal dragged the rear limbs.

a. What is the differential diagnosis for the paresis? How would you proceed to a definitive diagnosis?
b. What is the probable explanation for the fecal staining of the perineum?
c. In cases of spinal luxation, what factors determine a recommendation of cage rest or euthanasia?

Guinea Pigs

CASE 14

A graduate student in psychology submitted a 600-gm female guinea pig with a history of anorexia and death. Gross necropsy examination revealed atelectasis and consolidation of the entire left lung and congestion and focal abscessation of the right. The spleen was enlarged and had several white foci on its surface. The kidneys had white foci visible through the capsule.

a. The lesions in this guinea pig indicate a septicemia. The enlarged spleen with the necrotic foci is suggestive of which disease?
b. Is diarrhea a frequent finding in acute salmonellosis in research animals?
c. What prognosis can be given for colonies with endemic salmonellosis?

CASE 15

An adult Strain 13 guinea pig exhibited lethargy, anorexia, emaciation, incoordination, and excessive salivation. Necropsy examination revealed a suppurative otitis media and encephalitis. A hemolytic *Streptococcus* was cultured from the middle ear.

a. What is the probable species of the *Streptococcus*?
b. How would this organism reach the middle ear and brain?
c. What are the causes of ptyalism in guinea pigs?

CASE 16

Six adult Strain 2 guinea pigs had been given (SQ) 100 mg each guinea pig thyroglobulin. Five days later one animal exhibited labored breathing, dehydration, emaciation, and a purulent discharge around the nares and eyes. Gross necropsy examination revealed a bronchopneumonia with multiple intrathoracic, fibrinous adhesions. The gut was hyperemic and empty, the liver pale and friable, and the gall bladder distended with bile.

a. What bacterial organisms should be considered as etiologic agents in an acute bronchopneumonia in guinea pigs?
b. What immediate diagnostic procedures would you utilize to determine the causative agent?
c. What is the possible relationship between the experimental protocol and the bronchopneumonia?

CASE 17

An emaciated (365 gm) adult female guinea pig was submitted for necropsy examination. A diagnosis of scurvy had been given by the local practitioner. Clinical and gross necropsy examination of external structures and thoracic and abdominal cavities revealed no ectoparasites or gross lesions.

a. What structures had the prosector failed to examine?
b. Anorexia and emaciation are common in guinea pigs. Suggest several causes for this problem in pet guinea pigs.

CASE 18

One juvenile female guinea pig in a litter of three died suddenly. A skin scraping revealed the guinea pig fur mite. On gross necropsy examination the small intestine and stomach were hyperemic and distended with gas. Petechial hemorrhages were evident on the mucosal surface of the small and large intestines. Coliform organisms were isolated from the gut.

a. What is the genus and species of the guinea pig fur mite? Will this mite cause clinical disease?
b. How could these mites be eliminated from a pet guinea pig? From a guinea pig colony?
c. What factors would predispose a guinea pig to an acute, fatal enteropathy?

CASE 19

A guinea pig with pododermatitis was noticed in a breeding colony. The animal was given ketamine, and the foot lesion was palpated and incised with a sterile scalpel. Several attempts to locate an abscess were unsuccessful. The foot, bleeding profusely, was bandaged, and the animal was held 5 days for observation. The lower portion of the affected limb was later amputated.

a. What factors predispose to pododermatitis in guinea pigs?
b. If the inflammatory response in a pododermatitis is a chronic arthritis and not abscess formation, what treatment would you recommend?
c. Why might this guinea pig have become ketotic during the observation period?

CASE 20

A 620-gm guinea pig, which had delivered two young 5 days previously, was submitted for clinical examination. The right mammary gland was swollen and discolored.

a. How many mammary glands does a guinea pig have?
b. The guinea pig died 3 days after a penicillin injection. What is the probable cause of death?
c. What provisions should be made for the orphaned young?

CASE 21

A 4-year-old female guinea pig had a history of weight loss, depression, and sitting hunched up in the feed dish. There was a subcutaneous mass 2 × 3 cm in the right inguinal area. On necropsy examination the liver, spleen, and kidneys were congested and enlarged. Both the right hind foot and the right mammary gland were enlarged and abscessed. *Staphylococcus aureus* was cultured from both tissues. A homogeneous, eosinophilic material was observed histologically in the visceral organs.

a. What is the eosinophilic material? Is there any relationship between a chronic infection and the deposition of this material?
b. Mastitis is common in guinea pigs. Suggest a possible treatment for acute mastitis in the guinea pig.

CASE 22

An 11-month-old female guinea pig died during her first parturition. The hair coat was dull, and numerous, small, crawling, white insects were seen on the dark hairs. There was a small amount of blood around the external genitalia. Necropsy examination revealed the pubic symphysis to be separated one-half inch and subcutaneous and intramuscular hemorrhage around the symphysis. The left uterine horn contained no fetuses and was partially involuted. The right horn was nongravid.

a. At what weight are guinea pigs first bred? How may the late breeding age in this case be related to the dystocia?
b. Are lice insects? Can they survive off the host?
c. Are guinea pig lice blood suckers?

CASE 23

A professor brought his daughter's adult male guinea pig to a veterinary clinic. The animal sat hunched and refused to move. The stifle joints were swollen and firm. The professor said the guinea pig had been fed guinea pig food purchased at a local hobby shop.

a. What is the probable diagnosis of this animal's malady?
b. What therapeutic regimen would you recommend?
c. You suggest giving a drug in the water, but the guinea pig refuses to drink. How would you proceed with treatment?

Hamsters

CASE 24

A 4-week-old golden hamster was submitted for clinical examination. The animal had a "wet tail," yellow diarrhea, and a prolapsed rectum.

a. At what age are hamsters weaned?
b. How is proliferative ileitis prevented in young hamsters?
c. What cestode causes an enteritis in hamsters?

CASE 25

A golden hamster had not eaten for 4 days. The animal had blood at the anus and hair loss over the back and on both front legs. On necropsy examination the left adrenal gland was approximately 2 cm in diameter and had hemorrhaged into the abdominal cavity.

a. What diagnostic procedure should be undertaken to determine the agents involved in the dermatitis?
b. Do adrenal tumors occur in hamsters?
c. How are *Demodex* infections treated?

Gerbils

CASE 26

An adult gerbil on a diet of sunflower seeds exclusively (no water) was submitted moribund. Three days previously the animal had jumped off a scale and fallen 3 feet to the floor. The gerbil had a paralysis of the rear limbs.

a. What dietary deficiencies may occur with a sunflower seed diet?
b. What is the probable cause of the sudden onset paralysis?
c. How would this diet predispose to limb fractures?

CASE 27

Two 4-year-old gerbils were presented to the clinic. One gerbil had a soft, subcutaneous mass (1 × 2 cm) immediately posterior to the xiphoid process. An abscess was detected by aspiration. The abscess was drained and flushed, and the gerbil was given chloramphenicol and sent home. One month later the gerbil returned to the clinic with a firm mass in the same area. The second gerbil, a female, had a rough hair coat and a slight but fixed head tilt.

a. In the first case the abscess was a secondary problem. What neoplasm might be suspected? What gland is located on the gerbil's abdominal midline?
b. Why do sick animals often have rough hair coats?
c. Chloramphenicol and dexamethazone successfully stopped the progression of the torticollis. What is the therapeutic function of the steroid?

CASE 28

Two 17-day-old gerbils were presented for necropsy examination. Although the gerbils were cachexic, no gross lesions were present. The mother gerbil was in good health. The diet was a quality, pelleted laboratory chow. A large water bottle was attached to the side of the cage.

a. What questions would you have asked the client about the husbandry practices?
b. At what age are gerbils weaned?
c. What are the mechanics involved in operating a sipper-tube waterer?

CASE 29

An adult male gerbil died suddenly. Because of a leaky water bottle, the cage had been wet for a week. The gerbil had usually been fed rodent pellets, but 2 days prior to death carrots were substituted as the sole diet. The contents of the cecum and colon were bright orange, and the gastric and small intestinal contents were black.

a. Explain the unusual cecal content.
b. What husbandry factor(s) may have caused the death of the gerbil?

CASE 30

A family purchased a pair of 6-week-old gerbils. The family was hoping to breed the animals and requested from a veterinarian general information about breeding gerbils. The clients were also concerned about the gerbils' rough hair coats.

a. What causes gerbils to have rough hair coats?
b. How do you verify that the gerbils are a male and female pair?
c. They inquire about the type of cage to use, the best kind of feed, bedding, feeding and watering devices, and nesting boxes. What will the practitioner tell them?

Mice

CASE 31

A black New Zealand mouse with a rough hair coat and a severe head tilt was presented to a veterinary clinic. Small, white ectoparasites were seen on the hair shafts. On gross necropsy examination the spleen was noted to be approximately ten times normal size and the kidneys were pale brown. The tympanic bullae were grossly normal. No microbial organisms were cultured from the lungs, spleen, or middle ears. On histologic examination hematopoietic centers were observed in the liver and spleen. The glomerular basement membranes were thickened.

a. How do you account for the splenomegaly?
b. What is a possible cause for the anemia noted in this mouse?
c. How would you treat a colony of mice infested with fur mites?

CASE 32

Six, young male C3H mice, 4 alive and 2 dead, were emaciated and had bloody feces around the anus. All mice had either a catarrhal or hemorrhagic enteropathy and thickened colons. The livers of the animals were pale, and the spleens were enlarged. Culture of the colonic contents resulted in a heavy growth of *Proteus* spp and coliform-like organisms.

a. Why did some mice have pale viscera?
b. What pathologic processes might account for a thickened colon?
c. What etiologic agents might be suspected in this case?
d. How would you contain and resolve an outbreak of enteritis in a large mouse colony?

CASE 33

Two percent of the albino A strain mice in a commercial colony were scratching violently at their ears and neck. No mites or lice were found on the animals. Necropsy examination revealed unilateral and bilateral purulent otitis media among the mice examined.

a. Why didn't the mice with otitis media also have torticollis?
b. Why were the mice scratching?
c. What etiologic agents are commonly involved in otitis media in mice?
d. How can the condition be treated?

CASE 34

Three 4-week-old depressed and emaciated male mice had hunched postures, distended abdomens, and diarrhea. At necropsy the small intestines of the mice were pale and thickened and contained an increased amount of mucus and gas, particularly in the anterior small intestine.

a. What diagnostic procedures would you use to obtain a definitive diagnosis?
b. What protozoal agents might be involved in this case?
c. Why, generally, are weanling ani-

mals more susceptible to acute enteropathies than adults?

d. How can this protozoal condition be treated or controlled?

CASE 35

Ten 7 to 9 gm (3 wk old) white mice with stunted growth, rough hair coats, focal alopecia, and a yellowish diarrhea were submitted for necropsy examination. The large intestine contained fluid feces. Although morbidity in the colony was high, mortality was low.

a. What viral conditions in weanling mice cause diarrhea? How would you differentiate these conditions?
b. What agents or processes might cause focal alopecia in mice?
c. What murine viruses are carried subclinically in the mouse gut?

Rats

CASE 36

An old Sprague-Dawley rat had lost weight over the past 2 months. The animal had a rough hair coat and was sneezing. Necropsy examination of the rat's thorax revealed fibrinous adhesions and pulmonary abscessation and consolidation. A diagnosis of fibrinopurulent bronchopneumonia was given.

a. What bacterial organisms may cause a fibrinopurulent bronchopneumonia in rats?
b. What are the gross differences between pulmonary abscessation and bronchiectasis?
c. What diagnostic procedures are available to differentiate among the several organisms involved in murine suppurative pneumonia?

CASE 37

A young girl brought her 2 adult female pet rats to the clinic. Both rats had single, 2 cm, firm, movable subcutaneous masses in the axillary region. Fibrous connective tissue cells were seen on microscopic examination of the biopsy specimen.

a. What is the probable origin of the masses?
b. What histologic type of neoplasm occurs most often in the mammary gland of the rat?
c. Which anesthetic would you use if you wished to surgically remove the masses?

CASE 38

Three 250-gm male rats were presented to the clinic with 1 to 2 cm ulcerated lesions on the skin of the throat and neck. The rats infrequently scratched the ulcers. No ectoparasites were found on the fur. Several of the untreated lesions continued to enlarge, while others regressed and healed. Periodically new lesions would appear on different areas of the trunk.

a. How would you determine the etiologic agent in these rats?
b. What is the significance of the heavy growth of *Staphylococcus aureus* recovered from the lesion?
c. How would you treat the condition?

CASE 39

A colony of 40 Long-Evans rats experienced an outbreak of "squinting" eyes associated with a serous conjunctivitis. The eyelids were swollen, and exudate matted the fur about the eyes. After a week the outbreak subsided, although several rats had clouded corneas. Histopathologic examination of the harderian glands (behind the globe of the eye and around the optic nerve) revealed acute inflammation, necrosis, and squamous metaplasia of the lacrimal gland epithelium. *Staphylococcus aureus* in pure culture was recovered from the conjunctival sac of 3 rats. No organisms were cultured from the conjunctiva of several other affected animals.

a. What is the hair coat pattern of the Long-Evans rat?
b. What is the probable classification of the etiologic agent?
c. Why would corneal damage occur?

CASE 40

A junior high school student presented a litter of eight 12-day-old rats to your clinic with the complaint that 3 had lost their tails and 4 others had one to three annular constrictions on the proximal portion of the tails. Except for the tail problem, the young rats and dam appeared healthy. The girl realized the affected rats could not be helped, but she wanted specific advice for preventing the disease in future litters.

a. What is the postulated cause of the tail lesion?
b. How can this condition be prevented?
c. During which season is this condition most likely to occur?

Suggested Answers

These answers to the questions asked with the 40 clinical cases are intended to emphasize main points about the case and are not detailed discussions of each case or problem.

Rabbits

CASE 1

a. Provisions for drainage of urine, a soft, absorptive floor, and room for movement would have reduced the causative factors involved in this case of "sore hocks."
b. The ulcerative lesions should be cleaned with an antiseptic soap, rinsed, and dried. Topical and systemic antibiotics may be used as indicated. Soft, clean flooring should be placed in the box or cage.
c. Wetness reduces the thickness of the protective fur pad and promotes the growth of bacteria in skin irritated by urine and contaminated with feces.

CASE 2

a. Diagnostic procedures to determine the cause of pendulous abdomen, diarrhea, and cachexia in the rabbit would involve palpation of the intestines for distention, the liver for enlargement, and the stomach for a gastric hairball. The oral cavity should be examined for malocclusion and the feces for coccidial oocysts.
b. In hepatic coccidiosis the lesions are irregularly shaped (they follow the bile ducts) and have indistinct margins. A smear of a cut section of the liver or gallbladder would reveal oocysts on microscopic examination. *Taenia* lesions are usually rounded and have distinct margins. Larval cysts may be present in the peritoneal cavity. The numerous, small, white foci caused by *Francisella tularensis* may occur not only in the liver, but also in other viscera, including the spleen. Tularemia is rare in domestic rabbits, but more common in wild rabbits, especially in tick-infested areas.
c. *Eimeria stiedae* is the causative organism of hepatic coccidiosis in rabbits.

CASE 3

a. Specific causes of enteropathies in rabbits are difficult to determine. Microscopic examination of the feces of several rabbits or scrapings of intestinal mucosa may reveal coccidial oocysts. Culture of the feces for bacteria might be attempted, but the results are usually inconclusive. A histologic section of the affected intestine could be stained with Giemsa stain and examined for the intracellular *Bacillus piliformis* organism. Necropsy signs in enteropathies of rabbits are usually nonspecific.

b. Initial efforts to eradicate coccidiosis from a rabbitry would involve thorough mechanical scrubbing of equipment and premises, and installation of suspended, all wire cages, hopper feeders, and automatic waterers or water bottles with sipper tubes. Prophylactic feeding of sulfaquinoxaline supplemented feed to all rabbits for 1 month would reduce the prevalence of the carrier state in the colony. After elimination of the clinical disease and carrier state, the caging and equipment should be sterilized, and the rabbits should be returned to a nonsupplemented, quality pelleted rabbit chow.

c. The infertility may be due to temperature-induced thyroid hypofunction with resulting depression of spermatogenesis.

CASE 4

a. Torticollis, a consequence of otitis interna, is more common in older rabbits. Uncomplicated otitis media is rarely detected antemortem, except by radiographic examination.

b. Torticollis in a rabbit may be caused by verminous, protozoal, or mycotic encephalitis; neoplasia; trauma; poisoning, or a congenital abnormality, but the common etiology is a bacterial otitis interia.

c. Attempts to diagnose torticollis might include examination of the cerebrospinal fluid, the brain, and the middle ear.

CASE 5

a. *Pasteurella multocida* is the most common bacterium isolated from the nasal cavity of rabbits.

b. Rhinitis or "snuffles" in rabbits is difficult to cure. Systemic antibiotics will affect organisms in contact with the circulatory tract, but bacteria sequestered in the labyrinthine nasal passage can be reached only with a nasal aerosol containing a mucolytic compound, vasoconstrictive agent, and an antibiotic. Both systemic and topical treatments should be continued for 5 to 7 days.

c. The rabbit ear mite is *Psoroptes cuniculi*. Treatment involves cleaning the ear of scabs and crusts and applying an oil-based acaricide on two occasions approximately 7 days apart.

CASE 6

a. The renal lesion may be the manifestation of an interstitial nephritis caused by *Encephalitozoon cuniculi* (passed in the urine) or the arteriosclerosis-induced fibrosis associated with hypervitaminosis D in rabbits (vitamin D from cow's milk).

b. Pregnancy toxemia (ketosis) is the probable cause of death. Prevention of pregnancy toxemia or ketosis requires the elimination of obesity in breeding does by reducing intake or feeding high fiber feed and the *ad libitum* provision of a quality rabbit feed during pregnancy. Frequent, successive pregnancies also reduce obesity.

c. The rabbit's gestation period is between 28 and 34 days.

CASE 7

a. "Blue fur disease" in rabbits is caused by *Pseudomonas aerugi-*

nosa; the blue-green discoloration is due to the bacterial pigment pyocyanin. If the drinking water is contaminated, the water should be chlorinated or acidified.

b. Equipment must be thoroughly disinfected with a disinfectant effective against the specific *Pseudomonas* involved.

c. Radiation, corticosteroid administration, and other circumstances or agents which reduce resistance may precipitate clinical *Pseudomonas* infection.

CASE 8

a. Cachexia in rabbits may be due to malocclusion, a gastric hairball, or, less commonly, to a nutritional deficiency, chronic disease (coccidiosis, pasteurellosis, neoplasia), pain, advanced age, ectoparasitism, arteriosclerosis, or competition for or restricted access to feed or water.

b. Decreased food intake would lead to agalactia and dehydration, hypoglycemia, and death of the litter.

c. A gastric hairball may be removed surgically or loosened by the oral administration of mineral oil; however, the prognosis for recovery or resolution remains poor, because the animal may develop ketosis.

CASE 9

a. The lesion is probably in the genital tract, specifically the uterus. Uterine adenocarcinoma is a common neoplasm in older does. Metritis, abortion, and vaginitis may also occur.

b. Palpation for uterine tumors, radiography, and exfoliative cytology might be used to determine the presence of uterine neoplasia.

c. Clinical signs of uterine neoplasia in does include infertility, stillbirths, abortions, and a sanguineous vulvar discharge.

CASE 10

a. A decision must be made to treat or not treat an animal with multiple subcutaneous abscesses. Isolated abscesses may be treated by surgical excision and flushing with an antiseptic or antibiotic. A systemic antibiotic will protect against a septicemia resulting from surgical intervention.

b. The margins of the cutaneous lesion should be examined by hand lens for ectoparasites. A skin scraping may reveal ectoparasites or dermatophytes when examined microscopically. A sample of broken hair from the margin of the lesion can be cultured on dermatophyte growth media. The possibility of repeated abrasion of the skin on the cage or feeder should be considered as an alternative to an infectious process.

c. Rabbits with ectoparasitism may be treated topically with an acaricide dust or ointment, or the rabbit may be dipped into an insecticide solution.

CASE 11

a. Drooling or ptyalism in rabbits may be caused by malocclusion, or the dewlap may become wet and abraded if the rabbit feeds or drinks from crocks. A wet dewlap, which often is associated with a moist, bacterial dermatitis, may be prevented by correcting malocclusion or feeding from hopper feeders and watering from sipper-tube watering devices. Superficial, moist dermatitis is treated by shaving the hair over the lesion, cleansing the skin with an antiseptic soap, rinsing, and applying a topical antibiotic powder or ointment.

b. *Pasteurella multocida* may reach the conjunctival sac through the nasolacrimal duct or by transmission on the forepaws from the nares to the eye.
c. Conjunctivitis in rabbits may be treated with an ophthalmic ointment containing an antibiotic. Treatment of the conjunctival infection does not, of course, eliminate the organism in the nasal passage.

CASE 12

a. Palpation and needle aspiration will provide evidence that the mass is a suppurative or neoplastic process. A biopsy specimen examined histologically will indicate the type of neoplasm. If the tumor is of infectious or bacterial origin, the lesion can be cultured.
b. Fibromas in *Oryctolagus* are probably caused by a poxvirus resident in the cottontail population and transmitted to domestic rabbits by mosquitoes or fleas.
c. Poxvirus-induced fibromas usually regress after several weeks or months. Metastasis has been reported in young rabbits, but is uncommon. The amputation may have been premature.

CASE 13

a. A congenital abnormality involving the spine, coxofemoral junction, or limb bones would exhibit a familial inheritance pattern. A traumatic injury would have a history of sudden onset and an association with a fall, improper restraint, or other traumatic event. A definitive diagnosis would be established by radiographic, physical, and neurologic examination.
b. The rabbit dragged the rear quarters in the dirty cage and stained the perineum and ventrum.
c. The prognosis of posterior paresis in the rabbit depends on the extent of the damage to the spinal cord or nerve trunks. If the lesion is focal edema and inflammation, cage rest may restore motor function. If the cord or nerves are torn, recovery will not occur.

Guinea Pigs

CASE 14

a. Salmonellosis in laboratory animals causes generalized visceral congestion and focal necrosis.
b. Diarrhea may occur with *Salmonella* infection, but this sign is an inconsistent finding. Affected animals may have a soft, discolored stool.
c. Because of the epizootic and fatal consequences of a *Salmonella* outbreak, the persistence of the disease in a carrier state, and the public health significance, infected colonies should be destroyed, the facility should be thoroughly cleaned and disinfected, the animal care personnel should be examined, and the colony should be restocked with *Salmonella*-free animals.

CASE 15

a. *Streptococcus zooepidemicus* is the streptococcal organism commonly affecting guinea pigs.
b. The organism may enter the guinea pig via cutaneous wounds, respiratory aerosol, or the conjunctival, oral, or genital mucosa. The *Streptococcus* can reach the middle ear and brain through the circulatory system or by extension through the eustachian tube and middle and inner ear.
c. Ptyalism in guinea pigs has been associated with malocclusion, folic acid deficiency, fluorosis, scorbutus, and adrenocortical insufficiency.

CASE 16

a. Pneumonia in guinea pigs may be caused by a large number of bacteria, including *Bordetella bronchiseptica, Streptococcus pneumoniae, Streptococcus zooepidemicus,* and *Klebsiella pneumoniae.* Pleuritis is often due to *S. pneumoniae.*
b. A Gram or Wright's stained smear of the exudate should be examined to determine the morphology and staining reactions of the organism.
c. The organism may have been introduced with the needle, or the animal was stressed as a result of the experimental procedure. Many of the respiratory pathogens of the guinea pig are carried latently in the respiratory passages and initiate a disease process only if the host's resistance is lowered.

CASE 17

a. Emaciation in guinea pigs is often caused by malocclusion of the cheek teeth. The practitioner should also inquire about recent changes in the taste or composition of the animal's diet.
b. Causes of weight loss in a pet guinea pig include malocclusion, metastatic calcification, vitamin E deficiency, ectoparasitism, chronic renal disease, and anorexia induced by changes in the taste or composition of the feed or changes in the feeding and watering devices.

CASE 18

a. The guinea pig fur mite is *Chirodiscoides caviae.* In heavy infestations alopecia and pruritus may occur.
b. Individual animals may be dusted with a powder containing an insecticide, or a resin strip impregnated with the organophosphate dichlorvos may be placed in the cage or room for several 24 hour periods.
c. Acute, fatal enteropathy in a guinea pig may result from the administration of antibiotics that suppress Gram-positive flora, or from anorexia, dietary changes, or nonspecific stressors, which may alter the balance of intestinal microorganisms.

CASE 19

a. Pododermatitis in guinea pigs is most often encountered in heavy animals raised for long periods on rough or abrasive floors, such as wire.
b. Treatment of pododermatitis is difficult, since the inflammation is chronic and diffuse and relatively isolated from the vascular system. Staphylococci are frequently isolated from such lesions. Systemic antibiotics and surgical intervention might be attempted, but the prognosis for recovery is poor.
c. Ketosis in guinea pigs is a common sequela to anorexia. If for any reason (pain, septicemia), a guinea pig with a foot lesion should stop eating, ketosis could result.

CASE 20

a. A guinea pig has two (inguinal) mammary glands.
b. Penicillin, and most other antibiotics, when used in the guinea pig or hamster, may induce an alteration of the intestinal flora. This alteration is suspected to be an overgrowth of the Gram-negative flora and results in an enteritis or enterotoxemia.
c. Newborn guinea pigs are precocious and can eat softened, pelleted feed within a few hours of birth.

CASE 21

a. Generalized amyloidosis in guinea pigs has been associated with chronic staphylococcal infections.
b. Acute, bacterial mastitis may be treated with the application of heat to the affected area and the injection of cephaloridine or chloramphenicol for 5 to 7 days. Guinea pigs with mastitis are often depressed and refuse to eat, conditions related to toxemias.

CASE 22

a. Sows are first bred at approximately 500 gm and boars at 550 gm. Guinea pigs bred for the first time at 9 or more months may have a dystocia because of ossification and fusion of the pubic symphysis.
b. Lice are wingless insects and obligate ectoparasites, although they may survive for a few days off the host.
c. *Gliricola* and *Gyropus* are biting lice (mallophagans) and feed on tissue debris.

CASE 23

a. Pet guinea pigs with swollen joints usually have scorbutus or hypovitaminosis C.
b. The guinea pig must receive vitamin C, and the administration of vitamin drops is the easiest way to provide the vitamin. Each guinea pig should receive 10 to 15 mg/kg ascorbic acid per day. After the clinical deficiency has been resolved, the guinea pig should be fed a fresh, quality feed specifically compounded for guinea pigs. If the vitamin C content of the feed is suspect, then dietary supplementation with the vitamin is necessary.
c. Guinea pigs frequently refuse to eat or drink if the diet or feeders and waterers are changed. Vitamin drops administered with an eye dropper avoid the often self-destructive, fastidious habits of the guinea pig.

Hamsters

CASE 24

a. Hamsters are weaned at 3 to 4 weeks of age.
b. Proliferative ileitis is difficult to prevent in susceptible strains. Good sanitary measures, removal of nonspecific stressors, and the use of filter cage covers may reduce the severity of an outbreak.
c. *Hymenolepis nana* and possibly *H. diminuta* are cestodes that may be associated with catarrhal enteritis in hamsters.

CASE 25

a. A skin scraping from the margin of the skin lesion should be examined for *Demodex* spp and dermatophytes, although dermatophytosis has not been reported in the golden hamster. Alopecia in hamsters is also associated with endocrinologic imbalances.
b. Pheochromocytomas, cortical adenomas, and adenocarcinomas have been reported in golden hamsters. They are relatively common among the neoplasms of hamsters. Neoplasia may predispose an animal to demodicidosis.
c. *Demodex* infections may be treated with the daily application of a 1:5.5 ronnel concentrate and propylene glycol mixture for 5 weeks.

Gerbils

CASE 26

a. Sunflower seeds contain low levels of calcium and high levels of fat;

despite these deficiencies, gerbils provided free choice may prefer sunflower seeds to other diets.

b. The sudden onset paralysis probably resulted from a traumatic injury to the spine or limbs.

c. If the diet was, in fact, calcium deficient, the bones may have been more susceptible to fracture.

CASE 27

a. Cutaneous neoplasms have been associated with the ventral, midline sebaceous scent gland in the gerbil. These neoplasms are usually basal cell and squamous cell carcinomas.

b. Rough or dull hair coats in sick animals may be associated with anorexia, dehydration, elevated body temperature, wetting of the fur with saliva, or ectoparasitism and pruritus. The fur of healthy gerbils becomes matted in humid environments.

c. Torticollis indicates an inflammation of the inner ear. If an inflammatory process damages the vestibular apparatus of the inner ear, the animal may never regain a normal posture. Administration of a steroid may reduce the damage done by inflammation. If the torticollis is not severe, an affected animal may compensate and continue to eat and drink despite the head tilt.

CASE 28

a. Were the young gerbils able to reach and operate the water bottle? Were the young nursing the mother? Could the young animals gnaw the large, hard pellets? Did the young exhibit any signs of clinical disease?

b. Gerbils are weaned between 20 and 26 days of age. The gerbils in this case were preweanlings.

c. Water bottles with small, curved, or blocked sipper tubes may release water until the partial vacuum above the water prevents further flow of water out of and air into the bottle. The bottle may then appear full and functioning but will be inoperable for very small animals.

CASE 29

a. The orange material in the large intestine was partially digested carrot.

b. Sudden dietary changes or starvation in rodents may precipitate an acute gastroenteritis. This gerbil had a hemorrhagic gastroenteritis. The specific cause of the diet-induced gastritis is obscure.

CASE 30

a. Rough hair coats in gerbils are seen when the relative humidity is 50% or higher. Gerbils may also have rough hair coats if they are febrile, the water bottle leaks, or if the bedding is damp or contains resin.

b. Gerbils are sexed by noting the anogenital distance, which in adults is 5 mm for females and 10 mm for males. The female has a vulvar opening at the base of the urogenital papilla, whereas the male has a larger papilla. If the gerbils have identical arrangements of the anal and urogenital structures, they are not a male-female pair.

c. Gerbils should be provided with deep bedding and concealed nesting and hiding places within the cage. The bedding should not be abrasive or irritating. Gerbils should be fed a quality, pelleted rodent feed from a hopper feeder and watered from a water bottle. The well-known water conservation mechanism of the gerbil is a mechanism for survival in the desert and should not be used as a rationale for excluding water from the pet or laboratory gerbil. Nesting boxes may be simple metal squares 10 cm on a side.

Mice

CASE 31

a. New Zealand mice over 12 months of age have been reported to have autoantibodies against erythrocytes and thrombocytes, although the prevalence varies with population and substrain. New Zealand mice also develop antinuclear antibodies, glomerulosclerosis, and a lupus-like glomerulonephritis. The splenomegaly may be associated with active, extramedullary erythropoiesis.
b. If the mouse does have an autoimmune hemolytic anemia, this would explain the anemia. The mouse should be examined for hemorrhage into the intestinal tract.
c. The placement of small piece of a dichlorvos impregnated resin strip into the mouse cages will reduce the ectoparasite population. The strips should be placed in the cages for 24 hour periods every 10 days until the parasites are gone. Caution is required because organophosphate insecticides affect breeding and the cholinesterase level in mice.

CASE 32

a. Pale viscera indicate an anemia. As some of these mice had a hemorrhagic enteritis, this may account for the anemia.
b. Inflammation of the intestinal wall, with edema and inflammatory cell infiltration, will result in a thickened intestinal wall, as will hyperplasia of the intestinal epithelium. Hyperplastic colitis of mice is a fairly common disease caused by *Cotrobacter freundii.*
c. Enteric organisms that may be involved in enteropathies in mice include *Giardia, Hexamita, Citrobacter, Salmonella,* and *Pseudomonas.*
d. Rooms with affected mice should be isolated; the affected mice should be quarantined and treated or killed. Food, water, bedding, and the animal handlers should be examined for the causative agent. The facility should be thoroughly disinfected and restocked with known disease-free mice. Filter cage covers used in a mouse colony will reduce the transmission of the highly infectious disease agents.

CASE 33

a. Torticollis is caused by a lesion in the brain or inner ear. Uncomplicated otitis media does not, in itself, cause torticollis.
b. The otitis media may have caused the irritation of sensory nerve endings or trunks passing near the middle ear.
c. Otitis media in mice may be caused by *Mycoplasma* spp, *Pasteurella pneumontropica, Pseudomonas aeruginosa, Staphylococcus aureus,* and other respiratory pathogens of rodents.
d. Affected rodents can be given chloramphenicol injectable at 20 mg/kg for 5 to 7 days.

CASE 34

a. The intestinal content can be cultured on a specific medium intended for the isolation of intestinal pathogens, and a direct smear of duodenal content and the mucosa can be examined for protozoa.

b. The causative agent in this case is probably *Hexamita muris.*
c. Weanling animals lack both immunity and an existing intestinal flora; both deficiencies predispose to the establishment of an aberrant intestinal microbial population.
d. Dimetridazole at 0.1% aqueous solution used at weaning reduces mortality due to protozoal infections in mice.

CASE 35

a. Diarrhea in mice may be caused by a wide variety of organisms including the viruses of epizootic diarrhea of infant mice, mouse hepatitis virus, and Reovirus 3. EDIM affects weanling mice under 21 days of age. Mortality remains low, and the feces are pasty and yellow. Reovirus 3 infections are characterized by an oily or fatty diarrhea occurring in preweanling mice. On necropsy examination mice affected with Reo 3 may have pale foci in the liver and other viscera. Serologic or fluorescent antibody tests may be used to establish a definitive diagnosis.
b. Dermatophytoses, excessive grooming or barbering, biting, and abrasion on a cage surface will cause a focal alopecia in mice.
c. Murine viruses that may be shed in the feces include GD VII, mouse hepatitis virus, Reovirus 3, epizootic diarrhea of infant mice, ectromelia virus, K virus, and the minute virus of mice.

Rats

CASE 36

a. *Pasteurella pneumotropica, Corynebacterium kutscheri, Streptococcus pneumoniae*, and *Bordetella bronchiseptica* may cause a pneumonia in rats.
b. Abscesses contain pus and occupy parenchymal lung tissue. Bronchiectasis involves the accumulation of inflammatory debris, usually nonsuppurative, within the air passages of the lungs. *Mycoplasma* chronic pneumonia is often complicated by a bacterial infection.
c. Observation of a Gram-stained smear from the lesion, culture of the affected tissue on blood agar and *Mycoplasma* media, and histologic examination of the lung will provide an etiologic diagnosis.

CASE 37

a. Subcutaneous masses in rats are usually mammary neoplasms.
b. Mammary neoplasms in rats are most often benign fibroadenomas.
c. Methoxyflurane administered through a nose cone will provide adequate surgical anesthesia in the rat. An intraperitoneal injection of sodium pentobarbital would also provide satisfactory anesthesia.

CASE 38

a. The lesion should be cultured on sheep blood or mannitol salt agar, and the agent recovered, if any, should be injected into the skin of an unaffected but susceptible rat.
b. Consistent isolation of *S. aureus* from the lesion indicated that the bacteria are probably involved in the pathogenesis of the disease and are not wound contaminants.
c. Ulcerative staphylococcal dermatitis in rats may be treated with nail trimming and a sulfonamide or furacin-penicillin gel ointment or a systemic antibiotic such as lincomycin or chloramphenicol. Before treatment, the antibiotic sensitivity pattern of the organism should be established.

CASE 39

a. Long-Evans rats have dark hair over the head and dorsoanterior trunk and are often described as "hooded" rats.
b. The causative agent of sialodacryoadenitis in rats is a corona virus.
c. Corneal damage would result from the decreased or absent lacrimal secretion and from traumatization of the eyeball.

CASE 40

a. Ringtail in young rats results when the vessels in the tail respond with constriction to an ambient humidity of 20% or less. As the rat's tail serves as a temperature control device, the ischemic lesion must involve an aberration of the microcirculation regulatory mechanism.
b. Ringtail can be prevented by insuring an elevated ambient humidity in the rat breeding cage. Humidity can be raised by placing the young rats in a nest box supplied with adequate bedding and nesting material and by controlling the relative humidity of the room.
c. Ringtail occurs during cold seasons, when humidity is low in artifically heated rooms.

Index

Acariasis 73-76
Acepromazine maleate, 47
  dosage of, 45
  in guinea pigs, 51
  in rabbits, 50
Abortion in rabbits, 61
Adenocarcinoma, uterine, in rabbits, 97, 127
Alopecia, in gerbils, 65
  in guinea pigs, 63
  in hamsters, 64, 130, 138
  in mice, 67
  in rabbits, 60, 127
  in rats, 68
Ammonia, in disinfection, 4, 82
  in mycoplasmosis, 95
Amputation, pathologic, in mice, 67
Amyloidosis, in guinea pigs, 120, 138
  in hamsters, 101
Anemia, in mice, 68, 131, 140
  in rabbits, 62
  in rats, 69
Anesthetics, 46-52
  in gerbils, 51
  in guinea pigs, 50-51
  in hamsters, 51
  inhalant, 49
  injectable, 44-45, 48-49
  in mice, 51-52
  in rabbits, 49-50
  in rats, 52
Anorexia, 115-116
  in guinea pigs, 127, 128
  in rabbits, 61
Anthelmintics, 44
Antibiotics, 43, 46, 77, 129
Anticholinergics, 43, 47-48
Antimicrobials
  dosages of, 43-44
  toxicity of, 46
Arteriosclerosis in rabbits, 13, 118
Antidote for organophosphates, 45
Arthritis, in guinea pigs, 63
  in mice, 67
  in rats, 68
Ascorbic acid, for guinea pigs, 16-17, 18
  in hypovitaminosis C, 90
Atropinesterase in rabbits, 12, 48, 50
Atropine sulfate, 47-48
  atropinesterase and, 12, 48, 50
  dosage of, 45, 50, 51

Bacillus piliformis, 76-77
Bacterins, for *Bordetella*, 78
  for pasteurellosis, 103
  for tularemia, 114
Bedding, anesthesia and, 46, 48-49
  dermatitis and, 117
  for gerbils, 26
  for guinea pigs, 16
  for hamsters, 21
  for mice, 31
  for rats, 37
Bleeding, procedures for, 53
  urogenital. *See also* Hematuria
    in gerbils, 27
    in guinea pigs, 113
    in rabbits, 97, 127
    in rats, 111
Body weights, 42
  of gerbils, 26, 42
  of guinea pigs, 17, 42
  of hamsters, 23, 42
  of mice, 32, 42
  of rabbits, 7, 42
  of rats, 38, 42
Bordetella bronchiseptica, 77-78
  in guinea pigs, 9, 18, 77
  mycoplasmosis and, 95
  in rabbits, 9
Breeding programs, for gerbils, 26-27
  for guinea pigs, 17-18
  for hamsters, 22-23

Breeding programs (con't)
for mice, 32-34
for rabbits, 10-12
for rats, 37-39
Bruce effect, in mice, 33
in rats, 39

Cage cleaning, 3-4
for gerbils, 26
for guinea pigs, 16
for hamsters, 22
for mice, 31
for rabbits, 9-10
for rats, 37
Cage size, 2
for gerbils, 26
for guinea pigs, 16
for hamsters, 21-22
for mice, 31
for rabbits, 9
for rats, 37
Campylobacter spp, 108
Cannibalism, in gerbils, 27, 66
in guinea pigs, 18, 63
in hamsters, 23, 65
in mice, 33, 67
in rabbits, 12, 61-62
in rats, 38, 39, 69
Carbaryl, 45
Castration, by biting, 9
surgical, 52-53
Cephaloridine injectable, 43
Cestodiasis, 78-80
Cheek pouches of hamsters, 24
Cheyletiella parasitivorax, 73-75
Chirodiscoides caviae, 73-75
Chloramphenicol palmitate, 43
Chloramphenicol succinate, 43
Chloroform toxicity, 54, 68
Chlorpromazine hydrochloride, 46, 47
dosage of, 45
pentobarbital and, 48, 51, 52
Coccidiosis, 80-82, 125
Conjunctivitis, in gerbils, 66
in guinea pigs, 63
in hamsters, 64
in mice, 67
in rabbits, 61, 127
in rats, 68
Constipation in rabbits, 61
Coprophagy in rabbits, 12, 80
Corynebacterium kutscheri, 82-83
mycoplasmosis and, 95
Cottontail rabbits, myxofibroma viruses and, 97-98
taxonomy of, 7
Cutaneous swelling, in gerbils, 65, 130
in guinea pigs, 63
in hamsters, 64
in mice, 66
in rabbits, 60, 127
in rats, 68, 132
Cuterebriasis, 60

Demodex spp, 73-75, 130
Dental caries in hamsters, 24
Dermatitis, 117
in gerbils, 65
in guinea pigs, 62-63
in hamsters, 64
in mice, 66
in rabbits, 60
in rats, 68, 132
Dermatophytosis, 83-84
Diarrhea, in gerbils, 65
in guinea pigs, 63
in hamsters, 64
in mice, 67, 131
in nonspecific enteropathies, 117-118
in rabbits, 60, 125
in rats, 69
Diazepan, 45
Dichlorvos, 45, 75, 106
Diet, 2
for gerbils, 26
for guinea pigs, 16-17
for hamsters, 22
imbalances in, 118-119
for mice, 31-32
predisposing to disease, 3
for rabbits, 10, 119
for rats, 37
Dimetridazole powder, 43
Disease predisposition, 2-3
Disease prevention, in gerbils, 28
in guinea pigs, 18
in hamsters, 23
in mice, 34
in rabbits, 12
in rats, 39
Disinfectants, 3-4
for coccidia, 82
formaldehyde, 4-5, 75, 106
Drug dosages, 41-45
administration of, 42
factors affecting, 41
Dyspnea, in gerbils, 66
in guinea pigs, 63
in hamsters, 64
in mice, 67
in rabbits, 61
in rats, 69
Dystocia in guinea pigs, 17, 116, 129

Ectromelia (mouse pox) 54-55, 85-86
Eimeria spp, 80-82
Eimeria stiedae, 79, 80-82, 125
Embryonal nephroma, 98
Encephalitozon cuniculi, 86-87, 126
Encephalitozoonosis, 86-87, 126
Encephalomyelitis, mouse, 92-93
Enteropathies, mucoid, 94-95
nonspecific, 117-118
Eperythrozoon coccoides, anemia and, 68
lymphocytic choriomeningitis and, 91
mouse hepatitis and, 93
Epileptiform seizures in gerbils, 28, 47, 116
Epizootic diarrhea of infant mice (EDIM), 87-88, 132

Estrous cycle, of gerbils, 26-27
  of guinea pigs, 17-18
  of hamsters, 23
  of mice, 33
  of rabbits, 11
  of rats, 38
Ether, 49
  in guinea pigs, 50
  in hamsters, 51
  in mice, 52
  in rabbits, 49-50
Ethylhydrocupreine, 112
Euthanasia, 54

Feeders, 2
  cleaning, 4
  for guinea pigs, 16
  for rabbits, 7, 9, 82
Fentanyl-droperidol (Innovar-Vet), 46, 48
  dosage of, 44
  in guinea pigs, 50
  in hamsters, 51
  in rabbits, 50
  in rats, 52
Fighting, by gerbils, 26
  by guinea pigs, 15
  by hamsters, 21, 23
  by mice, 30-31
  by rabbits, 9
  by rats, 36
Formaldehyde fumigation, 4-5
  for lice, 106
  for mites, 75
Fostering. *See* Hand rearing
Fractures, 52
  in rabbits, 12-13, 120
Francisella tularensis, 114
  in hamsters, 23
  in rabbits, 12
Furazolidone, 43

Gerbils, 24-29
  acariasis in, 73-75
  alopecia in, 65
  anatomy of, 28
  anesthesia in, 51
  Bacillus piliformis in, 76-77
  bedding for, 26
  behavior of, 28
  bleeding of, 53
  body weight of, 26, 42
  breeding of, 26-27
  cage cleaning for, 26
  cannibalism in, 27, 66
  cestodiasis in, 78-80
  convulsions in, 66
  cutaneous swellings in, 65, 130
  dermatitis in, 65
  diarrhea in, 65
  disease prevention in, 28
  eating habits of, 26, 28
  epileptiform seizures in, 28, 47, 116
  estrous cycle of, 26-27
  feeding of, 26, 42
  fighting by, 26
  fostering of, 27
  gestation of, 27
  hand rearing of, 27
  housing for, 26
  Hymenolepis in, 28
  identification of, 2
  incoordination of, 66
  injection sites in, 42
  infertility in, 66
  life span of, 25
  lipemia in, 28
  litter desertion in, 66
  litter size in, 27
  marketing of, 25
  mites in, 73-76
  neoplasia in, 66, 98-99, 130
  nests of, 26, 27
  as pets, 25-26
  physiology of, 28
  pregnancy determination in, 27
  pseudopregnancy in, 27
  public health concerns with, 28
  radiation tolerance of, 28
  restraint of, 25
  rough hair coat in, 65, 130
  salmonellosis in, 28, 109-110
  seizures in, 28, 47, 116
  sexing of, 26
  sore nose in, 117
  sources of information on, 24-25
  sudden death in, 66, 130-131
  syndromes in, cutaneous, 65
    excitable tissue, 66
    gastrointestinal, 65
    miscellaneous, 66
    reproductive, 66
    respiratory, 66
  taxonomy of, 24-25
  torticollis in, 66, 130
  Tyzzer's disease in, 76-77
  urine of, 26
  varieties of, 24
  watering of, 26, 42
  weight loss in, 66, 130
Gestation period, in gerbils, 27
  in guinea pigs, 18-19
  in hamsters, 23
  in mice, 33
  in rabbits, 11
  in rats, 38-39
Gliricola porcelli, 105-106
Glomerulonephrosis, 101, 131
Griseofulvin, 43
Guinea pigs, 13-19
  abscesses in, 63
  acariasis in, 73-76
  alopecia in, 63
  amyloidosis in, 120, 138
  anatomy of, 18-19
  anesthesia for, 46, 47, 50-51
  bedding for, 16
  bleeding of, 53
  body weight of, 17, 42

Guinea pigs (con't)
Bordetella, in, 9, 18, 77-78
breeding of, 17-18
cage cleaning for, 16
cannibalism in, 18, 63
castration of, 52-53
Chirodiscoides in, 73-75
coccidiosis in, 80
conjunctivitis in, 63
convulsions in, 63
cutaneous swelling in, 63
dermatitis in, 62-63
dermatophytosis in, 83-85
diarrhea in, 63
disease prevention in, 18
dyspnea in, 63
dystocia in, 17, 116, 129
eating habits of, 16, 18
ectoparasitism in, 63
estrous cycle of, 17-18
feeding of, 16-17, 42
fighting by, 15
gestation of, 18-19
Gliricola porcelli in, 105-106
Gyropus ovalis in, 105-106
hand rearing of, 18
heat stress in, 88-89
histamine sensitivity of, 19
housing of, 15-16
hypovitaminosis C in, 89-90, 129
identification of, 2
incoördination in, 63
infertility in, 63
injection sites in, 42
Kurloff bodies in, 19
lice on, 105-106, 129
life span of, 14
litter desertion in, 63
litter size of, 18
malocclusion in, 116, 128
marketing of, 14
mastitis in, 116-117, 129
metastatic calcification in, 64, 118
mites in, 73-76, 128
nematodiasis in, 63
neoplasia in, 98
nephrosis in, 101-102
nests of, 18
otitis media in, 113
ovariectomy of, 53
paresis or paralysis of, 63
as pets, 14-15
pediculosis of, 105-106
physiology of, 18-19
pneumonia in, 63
pododermatitis in, 120, 128-129
pregnancy determination in, 18
pregnancy toxemia in, 106-107
prenatal mortality in, 63
pseudopregnancy in, 18
ptyalism in, 63, 128
public health concerns with, 18
restraint of, 14-15
salmonellosis in, 18, 109-110, 128
serologic tests for, 54-55
sexing of, 17
sources of information on, 14
Staphylococcus in, 120, 128-129
Streptococcus pneumoniae in, 111-112, 128
Streptococcus zooepidemicus in, 112-114, 128
sudden death in, 64
surgical procedures in, 52-53
syndromes in, cutaneous, 62-63
excitable tissues, 63
gastrointestinal, 63
miscellaneous, 64
reproductive, 63
respiratory, 63
taxonomy of, 13-14
torticollis in, 63
urine of, 4
vaccinations for, 18
varieties of, 14
vitamin C requirement for, 16-17, 18
watering of, 16-17, 42
weight loss in, 64, 128, 129
yersiniosis in, 64
Gyropus ovalis, 105-106

Haemodipsus ventricosis, 105-106
Hairball, gastric, 119-120, 126-127
Halogenated disinfectants, 4
Halothane, 49
in guinea pigs, 50
in mice, 52
in rabbits, 50
Hamsters, 19-24
abscesses in, 64
acariasis of, 73-75
alopecia in 64, 130, 138
amyloidosis in, 101
anatomy of, 24
anesthesia in, 51
Bacillus piliformis in, 76-77
bedding for, 21
bleeding of, 53
body weight of, 23, 42
breeding of, 22-23
cage cleaning for, 22
cannibalism in, 23, 65
cestodiasis in, 78-80
cheek pouches in, 24
convulsions in, 65
cutaneous swellings in, 64
Demodex in, 73-75
dental caries in, 24
dermatitis in, 64
diarrhea in, 64
disease prevention in, 23
ectoparasitism in, 64
estrous cycle of, 23
feeding of, 22
fighting by, 21, 23
fostering of, 23
gestation of, 23
hand rearing of, 23
hibernation of, 24
housing for, 21-22

Hymenolepis in, 23, 78-80, 130
identification of, 2
incoordination in, 65
infertility in, 64
injection sites in, 42
life span of, 20
litter desertion in, 65
litter size in, 23
lymphocytic choriomeningitis in, 23, 90-92
marketing of, 20
mites in, 73-76, 130
neoplasia in, 24, 98, 130
nephrosis in, 101-102
nests of, 23
physiology of, 24
polyuria in, 64
pregnancy determination in, 23
prenatal mortality in, 64-65
proliferative ileitis in, 23, 107-109, 129-130
pseudopregnancy in, 23
public health concerns with, 23
rectal prolapse in, 64
restraint of, 21
salmonellosis in, 23, 109-110
serologic tests for, 54-55
sexing, of, 22
sources of information on, 20
syndromes in, cutaneous, 64
excitable tissue, 65
gastrointestinal, 64
miscellaneous, 65
reproductive, 64-65
respiratory, 64
taxonomy of, 19-20
torticollis in, 65
tularemia in, 23
Tyzzer's disease in, 76-77
urine of, 4
varieties of, 20
watering of, 22, 42
weight loss in, 65
Hand rearing, of gerbils, 27
of guinea pigs, 18
of hamsters, 23
of mice, 34
of rabbits, 12
of rats, 39
Hares, 7
Heat stroke, 88-89
Hematuria, in guinea pigs, 113
in rabbits, apparent, 13
in rats, 111
Hepatitis, mouse, 93-94
Hexamitiasis, 131-132, 140-141
Hibernation in hamsters, 24
History protocol, 59-60
Housing, 1-2
for gerbils, 26
for guinea pigs, 15-16
for hamsters, 21-22
for mice, 31
for rabbits, 9-10
for rats, 37
Humidity, 2
for gerbils, 26, 131
for guinea pigs, 16
for hamsters, 22
for mice, 31
for rats, 37
Hutchburn in rabbits, 117
Hymenolepis spp, 78-80
in gerbils, 28
in hamsters, 23, 130
Hypnosis in rabbits, 50
Hypnotics, 48-49
Hypovitaminosis C, 89-90, 129
Hysterectomy in rabbit, 53

Identification, 2, 3
Ileitis, proliferative, 23, 107-109, 129-130
Incoordination, in gerbils, 66
in guinea pigs, 63
in hamsters, 65
in mice, 68
in rabbits, 62
in rats, 69
Infertility, in gerbils, 66
in guinea pigs, 63
in hamsters, 64
in mice, 33, 67
in rabbits, 61, 125
in rats, 69
Injection sites, 42
Innovar-Vet. *See* Fentanyl-droperidol
Insecticides, 45
Intubation, gastric, 42
tracheal, 50

Ketamine hydrochloride, 44
in gerbils, 51
in guinea pigs, 50
in hamsters, 20
in mice, 52
in rabbits, 50
in rats, 52
Ketosis, 106-107
Klossiella spp, 80
Kurloff bodies, 19

Lethal intestinal virus (LIVIM), 93
Lice, 105-106
Life span, of gerbils, 25
of guinea pigs, 14
of hamsters, 20
of mice, 30
of rabbits, 8
of rats, 36
Light intensity, 2
Liponyssus bacoti, 73-75
Listeriosis, 61
Litter desertion, in gerbils, 66
in guinea pigs, 63
in hamsters, 65
in mice, 67
in rabbits, 61-62
in rats, 69
Litter size, of gerbils, 27
of guinea pigs, 18

Litter size (con't)
of hamsters, 23
of mice, 33
of rabbits, 12
of rats, 39
Liver, rabbit, spots on, 76, 79, 81, 114
Lymphocytic choriomeningitis, 23, 34, 90-92
Lymphosarcoma in rabbits, 98

Malocclusion, 116, 128
Mammary neoplasia in rats, 99-100, 132
Marketing, of gerbils, 25
of guinea pigs, 14
of hamsters, 20
of mice, 30
of rabbits, 8
of rats, 36
Mastitis in guinea pigs, 116-117, 129
Menadione, 45
Metastatic calcification, 64, 118
Methoxyflurane, 49
in guinea pigs, 50
in hamsters, 51
in mice, 52
in rabbits, 50
Microsporum spp, 83-84
Mites, 73-76, 128,. 130
Moist dermatitis, 117
Mouse (Mice), 29-35
abscesses in, 66
acariasis of, 73-75
anatomy of, 34
anemia in, 68, 131, 140
anesthesia of, 46, 51-52
appendage necrosis in, 66
autoimmune disease in, 131
Bacillus piliformis in, 76-77
bedding for, 31
body weight of, 32, 42
bleeding of, 53
breeding of, 32-34
Bruce effect in, 33
cage cleaning for, 31
cannibalism in, 33, 67
cestodiasis in, 78-80
coccidiosis in, 80
conjunctivitis in, 67
cutaneous swelling in, 66
Corynebacterium in, 82-83
dermatitis in, 66
diarrhea in, 67
disease prevention in, 34
eating habits of, 31
ectoparasitism of, 66
ectromelia in, 54-55, 85-86
encephalomyelitis in, 92-93
epizootic diarrhea (EDIM) in, 87-88, 132
estrous cycle of, 33
feeding of, 31-32, 42
fighting by, 30-31
fostering of, 34
gestation in, 33
hand rearing of, 34
hepatitis (MHV) in, 93-94
hexamitiasis in, 132, 140-141
housing for, 31
identification of, 2
inbred strains of, 29, 34
infertility in, 67
injection sites in, 42
lice in, 105-106
life span of, 30
litter desertion in, 67
litter size in, 33
lymphocytic choriomeningitis in, 34, 90-92
marketing of, 30
mites in, 73-75
mouse pox in, 54-55, 85-86
mycoplasmosis in, 95-96, 131
Myobia musculi in 73-75
Myocoptes musculinus in, 73-75
nematodiasis in, 66
neoplasia in, 99
nests of, 33
orchidectomy of, 53
orchidectomy of, 53
otitis media in, 96, 104
pasteurellosis in, 104-105
pediculosis of, 105-106
pendulous abdomen in, 67, 131-132
as pets, 30-31
physiology of, 34
pinworms in, 66
pneumonia in, 67
polio in, 92-93
Polyplax in, 105-106
pregnancy determination in, 33
pseudopregnancy in, 33
Psorergates simplex in, 73-75
public health concerns in, 34
rabies vaccinations in, 34
Radfordia affinis in, 73-75
rectal prolapse in, 67
restraint of, 30
salmonellosis in, 34, 109-110
Sendai virus in, 110-111
serologic tests in, 54-55
sexing of, 32
sources of information on, 29-30
sudden death in, 68
surgical procedures in, 53
syndromes in, cutaneous, 66-67
excitable tissue, 67-68
gastrointestinal, 67
miscellaneous, 68
reproductive, 67
respiratory, 67
Syphacia spp in, 66
taxonomy of, 29
Tyzzer's disease in, 76-77
watering of, 31-32, 42
weight loss in, 68
Whitten effect in, 33
Mouse encephalomyelitis, 92-93
Mouse hepatitis (MHV), 93-94
Mouse polio, 92-93
Mouse pox (ectromelia), 54-55, 85-86
Mucoid enteropathy, 94-95

Multiceps serialis, 78-80
Mycoplasma pulmonis, 95-96, 104, 110
Mycoplasmosis, murine chronic, 95-96, 131
Myobia musculi, 73-75
Myocoptes musculinus, 73-75
Myxofibroma in rabbits, 97-98, 127
Myxomatosis, 97-98

Nasal discharge, in gerbils, 66
in guinea pigs, 63
in hamsters, 63
in mice, 67
in rabbits, 61, 126
in rats, 69
Nematodiasis, in guinea pigs, 63
in mice, 66
Neoplasia, in gerbils, 66, 98-99, 130
in guinea pigs, 98
in hamsters, 24, 98, 130
in mice, 99
in rabbits, 97-98, 127
in rats, 99-100, 132
Nephritis in rabbits, 87
Nephrosis, 101-102
Nests, of gerbils, 26, 27
of guinea pigs, 18
of hamsters, 23
of mice, 33
of rabbits, 12
of rats, 39
Neuroleptanalgesics, 48-49
Niclosamide, 44
Nitrate toxicity in rabbits, 117
Nonspecific enteropathy, 117-118
Nosematosis, 86-87, 126
Nutritional imbalances, 118-119

Oophorectomy in mouse, 53
Optochin, 112
Orchidectomy in mouse, 53
Orchitis in rabbits, 102
Organophosphates, anesthesia and, 49
poison antagonists for, 45
toxicity of, 68
Otitis media, in guinea pigs, 113
in mice, 96, 104
in rabbits, 102
in rats, 96, 112
Ovariectomy of guinea pig, 53
Ovulation in rabbits, 12
Oxytetracycline, 43
Oxytocin, 45

Paraldehyde, 49
in rabbits, 50
Paresis or paralysis, in gerbils, 130
in guinea pigs, 63
in mice, 68
in rabbits, 62
Pasteurella multocida, 102-104, 127
Pasteurella pneumotropica, 95, 104-105, 110
Pasteurellosis, bacterins for, 103
Pediculosis, 105-106
Pendulous abdomen, in mice, 67, 131-132
in rabbits, 61, 125
in rats, 69
Penicillin, dosage of, 43
toxicity of, 46, 77, 129
Pentobarbital sodium, 48-49
chlorpromazine hydrochloride and, 48, 51, 52
dosage of, 44
for euthanasia, 54
in gerbils, 51
in guinea pigs, 51
in hamsters, 51
in mice, 52
in rabbits, 49-50
in rats, 52
Peritonitis, in rabbits, 102
in rats, 112
Phenolic disinfectants, 4
Pinworms in mice, 66
Piperazine, 44
Pneumonia, in guinea pigs, 63
in mice, 67
in rabbits, 61
in rats, 69
Pododermatitis, ulcerative, in guinea pigs, 120, 128-129
in rabbits, 120, 125
Poison antagonists, 45
for antibiotics, 46
Polio in mice, 92-93
Polydipsia, in hamsters, 64
in rabbits, 61
Polyplax serrata, 93, 105-106
Polyplax spinulosa, 105-106
Porphyria in rats, 68
Pregnancy determination, in gerbils, 27
in guinea pigs, 18
in hamsters, 23
in mice, 33
in rabbits, 11
in rats, 39
Pregnancy toxemia, 106-107
Prenatal mortality, in gerbils, 66
in guinea pigs, 63
in hamsters, 64-65
in rabbits, 61
Proliferative ileitis, 23, 107-109, 129-130
Pseudomonas spp, in rabbits, 126
in water, 34
Pseudopregnancy, in gerbils, 27
in guinea pigs, 18
in hamsters, 23
in mice, 33
in rabbits, 9, 11-12
in rats, 39
Psorergates simplex, 73-75
Psoroptes cuniculi, 73-75, 126
Ptyalism, in guinea pigs, 63, 128
in mice, 68
in rabbits, 61, 127
in rats, 69
Public health concerns, with gerbils, 28
with guinea pigs, 18
with hamsters, 23
with mice, 34
with rabbits, 12
with rats, 39

Pyometra in rabbits, 102

Quaternary ammonium compounds, 4
Quellung reaction, 112

Rabbits, 7-13
abortions in, 61
abscesses in, 60
acariasis in, 62, 73-76
alopecia in, 60, 127
anatomy of, 11-12
anemia in, 62
anesthesia in, 46, 47, 49-50
anorexia in, 61
arteriosclerosis in, 13, 118
Bacillus piliformis in, 76-77
bleeding techniques in, 53
blood cells of, 13, 19
body weights of, 7, 42
breeding of, 10-12
cage cleaning for, 9-10
cannibalism in, 12, 61-62
castration of, 52-53
cestodiasis in, 78-80
Cheyletiella in, 73-75
coccidiosis in, 80-82, 125
conjunctivitis in, 61, 127
constipation in, 61
convulsions in, 62
coprophagy in, 12, 80
cutaneous swelling in, 60, 127
dermatitis in, 60
dermatophytosis in, 83-85
diarrhea in, 60, 125
disease prevention in, 12
dyspnea in, 61
ectoparasitism in, 60
Eimeria spp in, 80-82
encephalitozoonosis in, 86-87, 126
estrous cycle of, 11
feeding of, 7, 9, 10, 42, 82, 119
fighting by, 9
fostering of, 12
fractures in, 12-13, 120
Francisella tularensis in, 114-115
gestation in, 11
Haemodipsus in, 105-106
hairballs in, 119-120, 126-127
hand rearing of, 12
heat stress in, 88-89
housing of, 9-10
hutchburn in, 117
hypnosis of, 50
hysterectomy in, 53
identification of, 2
incoordination in, 62
infertility in, 61, 125
injection sites in, 42
lice in, 105-106
life span of, 8
listeriosis in, 61
litter desertion in, 61-62
litter size in, 12
liver spots in, 76, 79, 81, 114
malocclusion in, 116, 128
marketing of, 8
milk of, 12
mites in, 73-76
moist dermatitis in, 117
mucoid enteropathy in, 94-95
neoplasia in, 97-98, 127
nests of, 12
neutrophils of, 13
nitrate toxicity in, 117
nonspecific enteropathies in, 117-118
nosematosis in, 86-87, 126
orchitis in, 102
otitis media in, 102
ovulation in, 12
paresis or paralysis in, 62
pasteurellosis in, 102-104, 126, 127
pediculosis of, 105-106
pendulous abdomen in, 61, 125
peritonitis in, 102
as pets, 8
physiology of, 11-12
pneumonia in, 61
pododermatitis in, 120, 125
polydipsia in, 61
pregnancy determination in, 11
pregnancy toxemia in, 106-107, 126
prenatal mortality in, 61
Pseudomonas in, 126
pseudopregnancy in, 9, 11-12
Psoroptes in, 73-75
ptyalism in, 61, 127
public health concerns with, 12
pyometra in, 102
rabbit pox in, 60
restraint of, 8
rhinitis in, 61
salmonellosis in, 12, 109-110
sexing of, 10
skeleton of, 12-13
sore hocks in, 120, 125
sources of information on, 7-8
splayleg in, 119
Staphylococcus in, 60
sudden death in, 62, 126
surgical procedures in, 52-53
syndromes in, cutaneous, 60
excitable tissue, 62
gastrointestinal, 60-61
miscellaneous, 62
reproductive, 61-62
respiratory, 61
Taenia in, 78-80, 125
taxonomy of, 7
teeth of, 13
torticollis in, 62, 126
Treponema in, 115
Trichobezoars in, 119-120, 126-127
Trichophyton mentagrophytes in, 83-84
tularemia in, 12, 23, 114-115
Tyzzer's disease in, 76-77
ulcerative pododermatitis in, 120, 125
urine of, 4, 9, 10, 13
vaccinations in, 12, 97-98

vaginal discharge in, 61
varieties of, 7
venereal spirochetosis in, 115
vitamin D in, 13, 18
watering of, 10, 42
weight loss in, 62, 125, 126
wet dewlap in, 117, 127
Rabbit pox, 60
Rabies vaccination, in guinea pigs, 18
in mice, 34
in rabbits, 12
Radfordia affinis, 73-75
Radiography, 53-54
Rapid Plasma Reagin (RPR) test. 115
Rats, 35-40
alopecia in, 68
anatomy of, 39-40
anemia in, 69
anesthesia of, 46, 52
appendage necrosis in, 68
Bacillus piliformis in, 76-77
bedding for, 37
bleeding sites in, 53
body weight of, 38, 42
breeding of, 37-39
Bruce effect in, 39
cage cleaning for, 37
cannibalism in, 38, 39, 69
cestodiasis in, 78-80
conjunctivitis in, 68
Corynebacterium in, 82-83
cutaneous swelling in, 68, 132
dermatitis in, 68, 132
dermatophytosis in, 83-85
diarrhea in, 69
disease prevention in, 39
eating habits of, 37
estrous cycle of, 38
feeding of, 37, 42
fighting by, 36
gestation in, 38-39
hand rearing of, 39
housing of, 37
identification of, 2
infertility in, 69
injection sites in, 42
lice in, 105-106
life span of, 36
Liponyssus bacoti in, 73-75
litter desertion in, 69
litter size of, 39
mammary fibroadenoma in, 99-100
marketing of, 36
mites in, 73-75
mycoplasmosis in, 95-96
neoplasia in, 99-100, 132
nephrosis in, 101-102
nests of, 39
otitis media in, 96, 112
pediculosis of, 105-106
pendulous abdomen in, 69
peritonitis in, 112
as pets, 36
physiology of, 39-40
pneumonia in, 69
Polyplax in, 68, 105-106
porphyria in, 68
pregnancy determination in, 39
pseudopregnancy in, 39
ptyalism in, 69
public health concerns with, 39
restraint of, 36
ringtail in, 119, 133
salmonellosis in, 109-110
Sendai virus in, 110-111
serologic tests in, 54-55
sexing of, 37
sialodacryoadenitis in, 68, 132-133
sources of information on, 35-36
Staphylococcus in, 132
Streptococcus pneumoniae in, 111-112, 132
sudden death in, 69
syndromes in, cutaneous, 68
excitable tissue, 68
gastrointestinal, 69
miscellaneous, 69
reproductive, 69
respiratory, 69
taxonomy of, 35
torticollis in, 69
Tyzzer's disease in, 76-77
ulcerative dermatitis in, 132
varieties of, 35, 40
watering of, 37, 42
weight loss in, 69
Whitten effect in, 38
Rectal prolapse, in hamsters, 64
in mice, 67
Resin dichlorvos strips, 75, 106
Respiratory rate, of guinea pigs, 51
of hamsters, 51
of mice, 51
of rabbits, 49
of rats, 52
Restraint, of gerbils, 25
of guinea pigs, 14-15
of hamsters, 21
of mice, 30
of rabbits, 8
of rats, 36
Rhabdomyomatosis in guinea pigs, 98
Rhinitis in rabbits, 61
Ringtail in rats, 119, 133
Rough hair coat, in gerbils, 65, 130
in mice, 66
in rats, 68

Salmonellosis, 109-110
in gerbils, 28
in guinea pigs, 18, 128
in hamsters, 23
in mice, 34
in rabbits, 12, 109-110
Scorbutus, 89-90, 129
Sendai virus, 110-111
Sex determination, in gerbils, 26
in guinea pigs, 17
in hamsters, 22

Sex determination (con't)
in mice, 32
in rabbits, 10
in rats, 37
Sore hocks in rabbits, 120, 125
Sore nose in gerbils, 117
Spirochetosis, venereal, 115
Splayleg in rabbits, 119
Staphylococcus aureus, in guinea pigs, 120, 128-129
in rabbits, 60
in rats, 132
Streptobacillus moniliformis, 67
Streptococcus pneumoniae, 95, 111-112, 128, 132
Streptococcus zooepidemicus, 112-113, 128
Sudden death, in gerbils, 66, 130-131
in guinea pigs, 64
in hamsters, 65
in mice, 68
in rabbits, 62, 126
in rats, 69
Sulfamerazine, 43
Sulfaquinoxaline, 44
Sylvilagus spp, myxofibromas in, 97-98
taxonomy of 7
Syphacia spp, 66

Taenia pisiformis, 78-80, 125
Taenia taeniaformis, 78-80
Taxonomy, of gerbils, 24-25
of guinea pigs, 13-14
of hamsters, 19-20
of mice, 29
of rabbits, 7
of rats, 35
Teeth, rabbit, 13
Temperatures, body, 52
room, 2
for gerbils, 26
for guinea pigs, 16
for hamsters, 22
for mice, 31
for rabbits, 9
for rats, 37
Tetracycline, 44
Thiabendazole, 44
Thiamylal sodium, 45
Thiopental sodium, 45
Torticollis, in gerbils, 66, 130
in guinea pigs, 63
in hamsters, 65
in mice, 65, 131
in rabbits, 62, 126
in rats, 69
Toxemia, pregnancy, 106-107
Tranquilizers, 47
dosages of, 45
Treponema cuniculi, 115
Trichobezoars in rabbits, 119-120, 126-127
Trichofolliculoma in guinea pigs, 98
Trichophyton mentagrophytes, 83-84
Tularemia, 12, 23, 114-115
Tyzzer's disease, 76-77

Urine, blood in, 111, 113
of gerbils, 26
of guinea pigs, 4
of hamsters, 4
of mice, 93
of rabbits, 4, 9, 10, 13
Uterine adenocarcinoma in rabbits, 97, 127

Vaccinations, for myxofibroma viruses, 97-98
for rabies, 12, 18, 34
Vaginal discharge, in gerbils, 27
in hamsters, 23
in mice, 67
in rabbits, 61
Venereal spirochetosis, 115
Ventilation, 2
Vertebral luxation in rabbits, 12-13, 120
Vitamin A deficiency, 119
Vitamin C, 16-17, 18, 89-90
Vitamin D in rabbits, 13, 18
Vitamin E deficiency, 119

Warfarin, poison antagonists for, 45
Water, acidification of, 34
chlorination of, 34
consumption of, 42
for gerbils, 26, 42
for guinea pigs, 16-17, 42
for hamsters, 22, 42
for mice, 31-32, 42
for rabbits, 10, 42
for rats, 37, 42
Waterers, 2, 4, 130
for guinea pigs, 16
for hamsters, 22
for mice, 31
for rabbits, 7, 9
for rats, 37
Weight loss, in gerbils, 66, 130
in guinea pigs, 64, 128, 129
in hamsters, 65
in mice, 68
in rabbits, 62, 125, 126
in rats, 69
Wet dewlap in rabbits, 117, 127
Wet tail in hamsters, 23, 107-109, 129-130
Whitten effect, in mice, 33
in rats, 38
Wry neck. *See* Torticollis

Yersiniosis, 64